From Populations to Ecosystems

From Populations to Ecosystems

Theoretical Foundations for a New Ecological Synthesis

MICHEL LOREAU

PRINCETON UNIVERSITY PRESS

Princeton and Oxford

Published by Princeton University Press, 41 William Street, Princeton, New Jersey 08540
In the United Kingdom: Princeton University Press, 6 Oxford Street, Woodstock, Oxfordshire OX20 1TW

All Rights Reserved

Library of Congress Cataloging-in-Publication Data

Loreau, Michel.
    From populations to ecosystems : theoretical foundations for a new ecological synthesis / Michel Loreau.
        p.   cm. — (Monographs in population biology)
    Includes bibliographical references and index.
    ISBN 978-0-691-12269-4 (cl. : alk. paper) — ISBN 978-0-691-12270-0 (pb. : alk. paper)
    1. Ecology. 2. Biodiversity. 3. Population biology. I. Title.
    QH541.L67   2010
    577.8—dc22   2009052049

British Library Cataloging-in-Publication Data is available

This book has been composed in Times New Roman

Printed on acid-free paper. ∞

press.princeton.edu

Printed in the United States of America

1 3 5 7 9 10 8 6 4 2

10 9 8 7 6 5 4 3 2 1

Contents

Acknowledgments

This book is the result of a decade or so of work, much of which has been performed in collaboration with colleagues and students. These many collaborations have been instrumental in shaping my views and establishing the results summarized here. Fred Guichard, Bart Haegeman, Florence Hulot, Shawn Leroux, and Elisa Thébault gave useful comments on some of the chapters. Bob Holt and an anonymous reader reviewed the entire book and provided valuable suggestions to improve it. My book also benefited from intense scrutiny and passionate discussions from the students attending my course, Linking Community and Ecosystem Ecology, at McGill University. Brad Cardinale, Evan Economo, Kevin Gross, Helmut Hillebrand, Tony Ives, Tim Keitt, Tommy Lennartsson, Kathleen LoGiudice, Birte Matthiessen, Sam McNaughton, Patrik Nilsson, Rick Ostfeld, Jonathan Shurin, Chris Steiner, Dave Tilman, Juha Tuomi, Jasper van Ruijven, and my various coauthors kindly provided copyright permissions and/or electronic files of some of the figures reprinted or modified in the book. Amy Long's help in obtaining copyright permissions was particularly appreciated. I thank all of them gratefully. My warmest thanks go to Claire de Mazancourt, who not only read the whole book but also gave constant support in the form of love, understanding, occasional technical help, and scientific contribution to some of the new results presented in chapter 5.

Preface: On Unifying Approaches in Ecology

The vigorous growth of ecology from its origins in the late 19th century and early 20th century has been accompanied by its gradual fission into several distinct subdisciplines. The unified view of ecology that was present in a book like Lotka's *Elements of Physical Biology* (1925), which introduced many of the theoretical approaches that are still followed today, has given way to more specialized research programs. Although specialization is to some extent inevitable to make science more precise and predictive, it also creates problems. The conceptual frameworks of the various areas tend to become increasingly divergent over time, hampering communication across the discipline as a whole. This divergence is nowhere more apparent than between two of the major subdisciplines of ecology, i.e., community ecology and ecosystem ecology. These two subdisciplines have grown largely independently, each having its own concepts, theories, and methodologies. *Community ecology* is to a large extent an outgrowth of *population ecology*. It is mainly concerned with the dynamics, evolution, diversity, and complexity of the biological components of ecosystems; its starting point is the population and its interactions with other populations. *Ecosystem ecology* is mainly concerned with the functioning of the overall system composed of biological organisms and their abiotic environment; its starting point is the flow of matter or energy among functional compartments.

The separation of these subdisciplines is understandable insofar as they partly address issues at different hierarchical levels and different spatial and temporal scales. But it is harmful insofar as it is an obstacle to their unity and mutual enrichment. In the real world, populations and communities do not exist in isolation; they are parts of ecosystems, and, as such, they are subjected to constraints arising from ecosystem functioning, such as energy dissipation and nutrient cycling. These constraints can deeply alter species interactions and community properties, as we shall see in this book. On the other hand, ecosystems do not exist without their biological components; the latter impose their own constraints on ecosystem processes, as the disruptions generated by some biological invasions attest.

In a way, community ecology and ecosystem ecology provide two different perspectives on the same material reality. Real ecological systems are not either "communities" or "ecosystems," they are both one and the other at the same time—it is just the way we look at them that makes them communities or ecosystems. There is today a clear need for integration of the two subdisciplines. Reunifying these perspectives is an important scientific challenge, not only to progress our fundamental understanding of natural and managed ecosystems but also to allow human societies to develop appropriate responses to the global ecological crisis we are entering as a result of growing human environmental impacts on the Earth system. These impacts include destruction and fragmentation of natural habitats, pollution, climate change, overexploitation of biological resources, homogenization of biota, and biodiversity loss, and affect indistinctly the composition, dynamics, and functioning of ecosystems.

Ecosystem ecology, with its emphasis on higher-level complex systems, has also traditionally been divorced from evolutionary biology, with its emphasis on individual fitness and selection. Yet, ecosystem functioning is shaped by evolution, just as evolution is shaped by the constraints that arise from ecosystem functioning. Fully understanding the functioning of ecosystems and predicting their responses to environmental changes require incorporation of an evolutionary perspective, just as a complete theory of evolution cannot be achieved without consideration of ecosystem functioning.

The need for integration of population, community, ecosystem, and evolutionary ecology has been increasingly recognized during the last 20 years. There have been a number of attempts at doing so from a variety of perspectives, such as those provided by hierarchy theory (O'Neill et al. 1986), linking nutrient cycling and food webs (DeAngelis 1992), linking species and ecosystems (Jones and Lawton 1995), complex systems theory (Levin 1999; Solé and Bascompte 2006), linking biodiversity and ecosystem functioning (Kinzig et al. 2001; Loreau et al. 2002b), ecological stoichiometry across levels of biological organization (Sterner and Elser 2002), and the metabolic theory of ecology (Brown et al. 2004). Each of these perspectives has contributed to addressing part of the problem. But a broader synthesis of the various subdisciplines of ecology is still lacking.

But is such a synthesis possible, and what does it involve? There seems to be a proliferation of "unified theories" in ecology currently, which raises the question whether this is a feasible, or even desirable, enterprise. As a matter of fact, there are a number of different unifying approaches in ecology, each of which has both merits and limitations.

First are approaches that seek *generalizations across hierarchical levels of organization based on elementary physical and biochemical constraints.*

Although there is a limited set of fundamental physical and biochemical laws that constrain all biological and ecological systems, their number is sufficiently large that a variety of unifying approaches have been developed historically. These include approaches based on energy and thermodynamics (energy budgets, entropy, metabolism, temperature, etc.), materials and biochemistry (ecological stoichiometry), structure and topology (fractals, network theory, etc.), and dynamics (complex systems theory, catastrophe theory, self-organized criticality, etc.). Each of them has made important contributions to ecology and other sciences by uncovering similarities in patterns and processes among vastly different systems, scales, and hierarchical levels of organization. Their limitations match up to their success. By focusing on one specific constraint or set of constraints, they explore that part of reality that can be explained by these specific constraints. For instance, the metabolic theory of ecology is able to account for a range of macroecological patterns related to body size and temperature (Peters 1983; Brown et al. 2004). This suggests that elementary physiological constraints related to body size and temperature are powerful enough to govern a number of large-scale ecological patterns. All patterns in ecology, however, cannot be reduced to the influence of body size, temperature, or physiology. Ecology is the science of the complex interactions that bind the organisms and their environment together. Simple physical and physiological laws cannot be expected to provide a full understanding of these complex interactions since they express general constraints that are independent of this complexity. Other forces and constraints govern many other ecological patterns and processes over a wide range of spatial and temporal scales.

Simplifying theories that link previously unrelated properties within a hierarchical level represent a second type of unifying approach in ecology. The "unified neutral theory of biodiversity and biogeography" developed recently by Hubbell (2001) and others is an example of such an approach. This theory radically simplifies the description of communities by assuming all species to be equivalent, and thereby obtains simultaneous predictions for a range of community properties that were previously described by different models. Its strength is that it provides a consistent set of testable predictions that can serve as null hypotheses in community ecology. Its corresponding weakness is that it deliberately ignores the many demographic and functional traits that determine the ecology of species, and hence it remains confined within a specific description of reality.

A third type of unifying approach in ecology, and in science in general, consists in *merging the principles and perspectives of different disciplines* to create a synthesis that goes beyond the boundaries of each discipline. This is the approach that I champion in this book, based on previous efforts by DeAngelis (1992) and others. In many ways this approach is orthogonal, and hence potentially complementary, to the previous ones. Instead of seeking generalizations within or across hierarchical levels based on a specific perspective and a specific set of constraints, I seek to lay bridges between different perspectives and different sets of constraints. Such an approach also has limitations since it cannot pretend to build a single unified theory of ecology. But I accept this limitation happily. In fact, I believe that a monolithic unified theory of ecology is neither feasible nor desirable. Natural systems are too complex to be reducible to a unique description. My goal is to generate new principles, perspectives, and questions at the interface between different subdisciplines and thereby contribute to the emergence of a new ecological synthesis that transcends traditional boundaries. Working along these lines leads to a range of theories on different topics, although these theories obviously have to be compatible and complementary.

Accordingly, this book is the expression of an evolving research program. In this book, I synthesize a decade or so of theoretical work at the interface between population, community, ecosystem, and evolutionary ecology and set it within a coherent framework. Many questions have found answers in this work, but many new questions have also emerged from these answers. The book addresses both the answers and the questions.

I start by revisiting the basic principles that underlie the approaches of population and ecosystem ecology in chapter 1 and show how mass and energy budgets can be used as a basis for unification of these approaches in building community and ecosystem models. Chapter 2 provides a synthesis of the many existing theories of species coexistence, with a focus on their often overlooked implications for community-level functional properties. This chapter covers a relatively classical topic in ecology, but it does so from a slightly different perspective than usual, and the material it contains serves as a basis for several subsequent chapters. The functional consequences of species coexistence are further discussed in chapter 3, which examines how species diversity within a trophic level—a community property that is increasingly threatened by human environmental impacts—affects ecosystem functioning. The relationship between biodiversity and ecosystem functioning has emerged as a vibrant new research field during the last 15 years or so and has greatly contributed to fostering the integration of community ecology and ecosystem ecology that I champion. As a result, this theme runs through the whole book.

Many of the studies on this topic, however, have considered artificially simple systems with a single trophic level, thus ignoring the vast complexity of real ecosystems with their myriad trophic and nontrophic interactions between species. Chapter 4 provides some theoretical foundations to start to address this complexity. It analyzes the relationships between species interactions, biodiversity, and ecosystem functioning in food webs and interaction webs. There has been a long-standing debate in ecology over the relationships between the stability and diversity or complexity of ecosystems. This debate has resurfaced recently within the context of biodiversity and ecosystem functioning research. Chapter 5 provides new perspectives on this topic by explicitly distinguishing and linking stability properties at the population and ecosystem levels.

Up to chapter 5, only part of ecosystem functioning is considered—usually biomass and productivity of one or several trophic levels. Chapter 6 extends the scope of analysis to the overall functioning of ecosystems. It presents a coherent theory of material cycling and of its role in ecosystem functioning and shows how and when an indirect mutualism between autotrophs and heterotrophs arises from nutrient cycling at the ecosystem level. Chapter 7 puts all these results into a spatial context. It examines how biodiversity, ecosystem functioning, and the relationship between them are affected by spatial flows of nutrients and individuals across ecosystem boundaries, and the constraints that arise from these flows in metacommunities and metaecosystems. Last, I explore the evolutionary dimensions of ecosystem functioning in chapter 8. This final chapter discusses how natural selection leads to evolution of ecosystems and ecosystem properties and provides rigorous bases for the development of a much needed evolutionary ecosystem ecology.

Throughout the book I make use of relatively simple mathematical models to build and support my theories. I limit myself to their most salient, clearly interpretable results, leaving more detailed treatment to specialized publications. As a result of this choice, I have decided to leave aside some important issues, such as the interactions between several limiting nutrients and their consequences for ecosystem functioning. Although I take an active interest in this topic, its theoretical treatment requires more complex, stoichiometrically explicit models, which are beyond the scope of this book. Sterner and Elser's (2002) book provides a comprehensive overview of ecological stoichiometry, albeit from a more empirical perspective.

The book covers a wide range of topics. But these topics are strongly related to each other and follow a logical progression, from competitive communities, which are small subsets of ecosystems, to entire ecosystems, and from small scales to larger spatial and temporal scales. My book does not pretend to provide a comprehensive treatment of all the issues related to these topics, let alone a final resolution of these issues. Its main purpose is to show that merging the principles of population, community, evolutionary, and ecosystem ecology opens up new ways to look at reality, thereby offering new insights into a wide range of key issues in ecology and new theoretical weapons to face the mounting ecological crisis. I am convinced that we need a more synthetic ecology and hence a more synthetic ecological theory. If this book contributes to spread this conviction, I shall regard it as successful.

From Populations to Ecosystems

CHAPTER 1

Population and Ecosystem Approaches in Ecology

Building a theory that merges population, community, and ecosystem ecology requires at the very least that the fundamental descriptions of reality provided by the various subdisciplines be compatible with each other. But meeting this basic requirement is far from being a trivial issue given the widely different conceptual foundations and formalisms used by population and community ecology on the one hand and by ecosystem ecology on the other. In this introductory chapter, I first briefly revisit the foundations and formalisms of the population and ecosystem approaches in ecology. I then show how mass and energy budgets can bridge the gap between them. Last, I present a minimal ecosystem model to illustrate how an approach based on mass and energy budgets can be used to build simple models that combine the flexibility of demographic models and the physical realism of ecosystem models. The approach developed in this chapter will be the basis for most of the models presented in the rest of the book.

THE FORMALISM OF POPULATION DYNAMICS: EXPONENTIAL AND DENSITY-DEPENDENT GROWTH

A population is a set of organisms from the same biological species in a given area. Since all individuals belonging to the same species are very similar to each other when considered over a whole life cycle, classical approaches to population ecology ignore variability among individuals and assume that these are identical. As a consequence, population dynamics focuses on changes in the number or density of individuals that make up the population. Thus, population ecology fundamentally has a demographic approach to reality, in which the basic unit of measurement is the individual.

Population dynamics is implicitly or explicitly based on the following balance equation, which tracks the fate of individuals from time t to time $t + 1$:

$$N_{t+1} = N_t + B + I - D - E. \tag{1.1}$$

In this equation, N_t is the number of individuals at time t, and B, I, D, and E are the numbers of births, immigrants, deaths, and emigrants, respectively, during the time interval from t to $t + 1$. The time unit is arbitrary; it may be a day, a year, or a generation, depending on the kind of organisms considered. This demographic balance equation simply states that the population at time $t + 1$ is the population at time t, plus the individuals that have been added to the population by birth or immigration, minus the individuals that have been removed from the population by death or emigration.

In the simplest case, assume a closed population (no immigration or emigration), a constant environment, and density-independent growth; i.e., the per capita demographic parameters are independent of population density. In this case, $I = E = 0$, $B = bN_t$, and $D = dN_t$, where b and d are constant per capita birth and death rates, respectively. Equation (1.1) then reduces to the familiar equation

$$N_{t+1} = N_t + bN_t - dN_t = \lambda N_t, \tag{1.2}$$

where $\lambda = 1 + b - d$ is the finite rate of increase of the population.

This equation says that population size is multiplied by a factor λ during each time unit. Starting from $t = 0$ and iterating the process over t time units yields

$$N_t = N_0 \lambda^t. \tag{1.3}$$

Thus, the population is predicted to grow geometrically at a rate λ per time unit.

An identical prediction is obtained assuming that demographic processes are continuous instead of discrete in time, which leads to the following differential equation:

$$\frac{dN}{dt} = \beta N - \delta N = rN, \tag{1.4}$$

where β, δ, and r are instantaneous per capita rates of birth, death, and population growth, respectively. This equation can be integrated to give

$$N(t) = N_0 e^{rt}, \tag{1.5}$$

which is identical to equation (1.3) with $\lambda = e^r$.

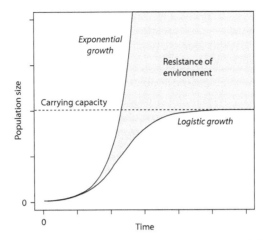

FIGURE 1.1. Exponential and logistic population growth. Populations tend to grow exponentially in the absence of environmental constraints, but logistically up to a carrying capacity when resources are finite. The difference between the two growth curves (gray area) can be interpreted as the resistance of the environment to unlimited growth. The carrying capacity, however, is a phenomenological abstraction that stands for a wealth of undefined ecological factors.

This fundamental equation of population dynamics, which is also known as Malthus's (1798) law, describes the inherent capacity of all organisms for exponential or geometric population growth (figure 1.1). *Exponential population growth* is a remarkably robust property as long as population processes are density-independent. It holds irrespective of spatial and temporal variations in demographic processes and population structure. If demographic processes vary in space or time, the finite and instantaneous population growth rates λ and r are simply replaced by appropriate spatial or temporal averages in equations (1.3) and (1.5). If age or stage structure is explicitly considered, the finite population growth rate λ is obtained from the projection matrix that describes transitions of individuals among age or stage classes (Caswell 1989). Exponential growth has been shown in numerous laboratory and natural populations under conditions of unlimited resource availability or low population density. The global human population itself is roughly experiencing exponential growth. More details on this topic can be found in theoretical ecology textbooks (e.g., Case 2000).

The propensity of populations to grow exponentially is an expression of the autocatalytic nature of biological systems and represents a fundamental source of instability in ecological systems. Basically, all organisms multiply as much as they can—until something prevents them from continuing to do so. And that is where ecology comes into play. All organisms are embedded in a complex web of interactions with their environment, which includes other organisms as well as abiotic factors. As populations grow, they modify their own environment through these multiple interactions, which feeds back on their capacity to grow further.

Classical population ecology makes the simplest possible assumption regarding these environmental feedbacks: it assumes that they can be reduced to a dependence of demographic processes on the population's own density. As the population grows, it progressively exhausts resources such as space, food, and nutrients, and as a result it decreases its potential to grow further. This convenient assumption eliminates the need to consider the complex web of interactions that organisms maintain with their environment and focuses on their net effect on the population variable under consideration. *Density dependence* is formally defined as a dependence of the per capita population growth rate on population density. In the continuous formalism of equation (1.4), which is the formalism that I shall use in most of this book, density dependence is expressed as

$$\frac{dN}{Ndt} = f(N), \tag{1.6}$$

with $f'(N) \leq 0$; i.e., the per capita growth rate monotonically decreases as population size increases.

The simplest form for the density-dependence function $f(N)$ is a linear form, which yields the classical *logistic equation* proposed by Verhulst (1838):

$$\frac{dN}{dt} = rN\left(1 - \frac{N}{K}\right). \tag{1.7}$$

In this equation, r, which is known as the intrinsic rate of natural increase, represents the maximum instantaneous population growth rate when population density is very low (close to zero), and K is known as the carrying capacity.

The logistic equation predicts a sigmoid growth pattern with a nearly exponential growth at low population size and a nearly exponential approach to a stable equilibrium population size equal to the carrying capacity (figure 1.1). This can be seen easily by noting that when population size is very small compared with the carrying capacity ($N \ll K$), the term in parentheses in equation (1.7) vanishes, and equation (1.4) describing exponential growth is recovered. On the other hand, when N approaches K, a first-order Taylor expansion of the right-hand side of equation (1.7) around K yields

$$\frac{dn}{dt} = -rn, \qquad (1.8)$$

where $n = N - K$ is a perturbation from the equilibrium value K. Thus, the logistic equation predicts an exponential decline of perturbations in the vicinity of the carrying capacity at the same rate as the exponential growth of population size at low density. In other words, density dependence stabilizes the population by counteracting its inherent tendency toward exponential growth and instability.

Logistic growth has been shown in numerous populations, especially in the laboratory under resource limitation. The reason why the logistic equation works so well under controlled laboratory conditions is simple: the linear density-dependence function in the logistic equation may be viewed as a first-order approximation to any form of density dependence. Logistic growth, however, is much less robust than density-independent exponential growth. Departures from the implicit assumptions of continuous demographic processes, constant environmental conditions, instantaneous operation of density dependence, and lack of population structure, can lead to periodic or chaotic population dynamics under logistic growth. These dynamical behaviors are qualitatively different from the stable equilibrium point predicted by the classical model. Again, more details on this topic can be found in theoretical ecology textbooks (e.g., Case 2000).

A more fundamental problem—from the perspective developed in this book—is that the density dependence included in the logistic equation in the form of the carrying capacity is a phenomenological abstraction. Parameter K is a condensed substitute for a wealth of factors and interactions that limit population growth, such as resources, competitors, mutualists, predators, parasites, and diseases. It is not even possible to disentangle the contributions of birth and death processes to density dependence in equation (1.7) since these are lumped into the parameters r and K.

Despite these limitations, the logistic equation has served as a basis for much of theoretical community ecology. The famous Lotka–Volterra models for interspecific competition or mutualism are direct extensions of the logistic equation in which the density-dependence function $f(N)$ in equation (1.6) is simply expanded to become a linear function of the population sizes of other interacting species. The classical Lotka–Volterra model for predation does not include direct density dependence but is built on the same principle; i.e., per capita growth rates are linear functions of population sizes.

Although many refinements and developments have been added to the theoretical corpus of community ecology, community ecology is largely an outgrowth of population ecology in its conceptual and methodological foundations. Most dynamical models in community ecology are based on a demographic approach that implicitly takes into account demographic balance constraints of the kind encapsulated in equation (1.1), but they ignore explicit physical constraints such as mass and energy balance (although there are exceptions, of course). As a result, community ecology has a strong focus on the structure, dynamics, and complexity of ecological systems, but it generally does not consider their overall functioning.

THE FORMALISM OF ECOSYSTEM FUNCTIONING: MASS AND ENERGY FLOWS

Ecosystem ecology does not have a simple fundamental law equivalent to the Malthusian law of exponential growth in population dynamics. Consequently, the approaches developed to model ecosystems have been somewhat more variable than in population ecology. The simplest and most common approach, however, has been that of compartmental modeling, which was pioneered by Lotka (1925). Ecosystem ecology is mainly concerned with the stocks and fluxes of materials or energy through the system as a whole, and this is explicitly what compartmental models represent. A *compartmental model* describes a set of compartments, the size of which is measured by the stock of materials or energy they contain, which are connected by fluxes of materials or energy. *Mass or energy balance* is explicitly taken into account in the description of these fluxes.

The basic building block of these models is a single-compartment model open to material or energy exchanges with the outside world. As an example borrowed from DeAngelis (1992), take a water body with a constant volume V that contains a solute of concentration C, and through which water flows at a constant rate q' per unit time, and let C_I be the solute concentration in the inflowing water. The principle of conservation of mass states that the rate of change of the mass of solute in the compartment equals the rate at which mass enters that compartment minus the rate at which mass leaves that compartment. Since the mass of solute in the compartment is CV, this principle is expressed in the following dynamical equation:

$$\frac{d(CV)}{dt} = q'C_I - q'C. \tag{1.9}$$

Dividing both sides by the constant volume V and rescaling the water flow rate as $q = q'/V$ yields

$$\frac{dC}{dt} = q(C_I - C), \tag{1.10}$$

which has the solution

$$C(t) = C_I + (C_0 - C_I)e^{-qt}. \tag{1.11}$$

This solution shows that the solute concentration in the water body tends asymptotically to the concentration in the inflowing water (the second term on the right-hand-side tends to zero as time goes to infinity, which leaves $C = C_I$) and that the deviation between the initial (C_0) and final (C_I) concentrations declines exponentially with time at a rate q. Thus, this system smoothly approaches a stable equilibrium concentration set by the inflowing water, at a rate governed by water flow. The water flow rate q sets the characteristic time of the system. It measures the rate at which the system approaches its equilibrium, which is one common measure of resilience (DeAngelis 1992). It also determines the mean residence time of the solute in the compartment, also called the turnover time of the system, which is obtained as the ratio of the equilibrium mass of solute ($C_I V$) over the equilibrium mass flow of solute ($q' C_I = q C_I V$), i.e., $1/q$.

This single-compartment model can easily be generalized to an arbitrary number of compartments coupled by material or energy flows. Take, for example, an ecosystem with two compartments 1 and 2, in which compartment 1 (say, plants) receives an input of a material such as carbon, part of the carbon contained in compartment 1 is transferred to compartment 2 (say, animal consumers) through some trophic interaction, and both compartments lose carbon to the external world through respiration or some other interaction (figure 1.2). Call X_i the carbon stock of compartment i, f_{ij} the rate at which a unit of carbon is transferred from i to j, with 0 standing for the external world, and I_{01} the input of carbon to compartment 1 per unit time. The principle of conservation of mass then yields the following system of differential equations:

$$\begin{aligned}
\frac{dX_1}{dt} &= I_{01} - f_{10}X_1 - f_{12}X_1, \\
\frac{dX_2}{dt} &= f_{12}X_1 - f_{20}X_2.
\end{aligned} \tag{1.12}$$

This system can be rewritten in matrix form as

$$\frac{d\mathbf{X}}{dt} = \mathbf{FX} + \mathbf{I}, \tag{1.13}$$

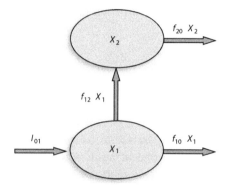

FIGURE 1.2. An abstract two-compartment ecosystem model. Circles represent energy or material stocks, while arrows represent energy or material flows.

where

$$\mathbf{X} = \begin{pmatrix} X_1 \\ X_2 \end{pmatrix}, \qquad \mathbf{F} = \begin{pmatrix} -f_{10} - f_{12} & 0 \\ f_{12} & -f_{20} \end{pmatrix}, \qquad \mathbf{I} = \begin{pmatrix} I_{01} \\ 0 \end{pmatrix}.$$

It is not difficult to show that this coupled system has similar properties as the previous single-compartment system. The equilibrium values (which will be denoted by an asterisk in this book following common usage) of the carbon stocks of the two compartments are easily obtained by setting the time derivatives equal to zero in equations (1.12) and solving for X_1^* and X_2^*:

$$X_1^* = \frac{I_{01}}{f_{10} + f_{12}},$$

$$X_2^* = \frac{I_{01} f_{12}}{(f_{10} + f_{12}) f_{20}}.$$

(1.14)

These equilibrium stocks are proportional to the carbon input into the system as before. The stability of the system is now governed by the eigenvalues of matrix $\mathbf{F}$, which contains the rate constants of carbon flows (except for the carbon input, which is independent of the system's dynamics). These eigenvalues can be shown to be both real and negative, thereby ensuring that the equilibrium is asymptotically stable (May 1973; Puccia and Levins 1985).

In contrast to population dynamical models, compartmental models used to describe mass and energy flows in ecosystems seem to be particularly stable and well behaved. This is, however, a consequence of the implicit or explicit assumptions about the physical constraints that govern these systems. For instance, the above single-compartment model is really a physical model based on the assumption that water flow drives the dynamics of the solute. The two-compartment ecosystem model assumes that carbon flows are either constant (driven from outside) or linear functions of internal carbon stocks, which erases the complexity of biological interactions that might be involved in these transfers. Thus, the smooth behavior of ecosystem models is often a consequence of the perspective adopted by ecosystem modelers. Ecosystem ecology has traditionally been concerned with predictable whole-system functional processes, ignoring much of the diversity and dynamical complexity of the organisms that constitute them. Some have even argued that linearity is an intrinsic property of ecosystem processes (Patten 1975), but this is a viewpoint that cannot be taken at face value. We shall return to this issue of the stability and predictability of ecosystem processes with new insights derived from more rigorous theory in chapter 5.

MASS AND ENERGY BUDGETS AS A BASIS FOR UNIFYING POPULATION AND ECOSYSTEM APPROACHES

The demographic and functional perspectives offered by population and ecosystem ecology are rooted in different concepts and principles. But, clearly, population dynamics has to be compatible with the physical principles of conservation of mass and energy, just as ecosystem functioning has to be compatible with the demographic law of exponential growth. How, then, can we lay a bridge between these two approaches?

Ecosystem ecology is essentially a physiology of ecological systems. It analyzes the functioning of an ecosystem in ways similar to those of physiology for individual organisms. In particular, ecosystem ecology and ecophysiology share the concepts of mass and energy budgets as tools for understanding the acquisition, allocation, and disposal of materials and energy in the metabolism and life cycle of both organisms and ecosystems. On the other hand, growth and reproduction are the two processes at the individual level that are responsible for population growth, and these processes place high demands on energy and materials in the metabolism of individual organisms. Thus, the unification of population and ecosystem approaches should be rooted in the ecophysiology of organisms, in particular, in the constraints that govern the acquisition, allocation, and disposal of materials and energy.

The realization that generic physiological constraints should act across all levels of biological organization is the basis for the recent development of two successful areas of ecology, i.e., ecological stoichiometry and metabolic theory. Ecological stoichiometry studies the balance among the chemical elements that make up living organisms (in particular, carbon, nitrogen, and phosphorus) and the constraints it generates for the functioning of biological systems, from cells to ecosystems (Sterner and Elser 2002). It is based on simple, fundamental physical and physiological laws, i.e., the conservation of mass and the homeostasis of living beings. The metabolic theory of ecology (Brown et al. 2004) is a quantitative theory that seeks to explain how metabolism varies with body size and temperature (essentially at macroecological scales) and constrains ecological processes at all levels of organization, from individuals to ecosystems. It is also based on simple constraints that govern the allocation of energy and materials in organisms.

The processing of energy and materials by individual organisms similarly constrains demographic processes at the population level. In principle, it should be possible to trace demography back to the *mass and energy budgets* of the individual organisms that make up the population. Energy budgets have been widely studied, especially in animals (Petrusewicz and Macfadyen 1970; Kooijman 2000). I am not so much interested here in the details of these budgets as in establishing simple approximate relationships between the parameters of classical population models and the components of these budgets. There have been several attempts to do so in the past (see Yodzis and Innes 1992 and references therein).

Here I start with a typical animal energy budget, which has the form (Petrusewicz and Macfadyen 1970)

$$C = A + Eg = P + R + Ex + Eg. \tag{1.15}$$

The amount of energy ingested by the organism (consumption, C) during some time period can be divided into a part that is assimilated (assimilation, A) and a part that is not. Nonassimilated energy is rejected without being digested (egestion, Eg) and corresponds to feces in animals. Assimilated energy is used for production of new tissues (growth) and new individuals (reproduction) (combined in production, P), respiration (R), and excretion of urine or other metabolic products (Ex). These elements of the energy budget are commonly used to define three measures of an organism's energetic efficiency: assimilation efficiency (A/C), gross production efficiency (P/C), and net production efficiency (P/A).

All the elements of the energy budget do not respond in the same way to increased food consumption (C). Part of the energy dissipated in respiration, and to a lesser extent in excretion and egestion, is used for basal metabolism, i.e., for the fixed energy costs of a living organism. Therefore, it is essentially constant. When food consumption is insufficient to match basal metabolism, the organism loses weight and eventually dies. Its production is then negative. When food consumption is greater than basal metabolism, the excess energy is used in positive production and active metabolism. These then increase roughly in proportion to consumption above the threshold consumption necessary to compensate for basal metabolism (Warren 1971).

These empirical relationships can be expressed mathematically as follows. Call B biomass and μ the mass-specific basal metabolic rate. Then if consumption is insufficient to match basal metabolism ($C \leq \mu B$), $R + Ex + Eg = \mu B$, and by the conservation equation (1.15),

$$P = C - \mu B < 0. \tag{1.16}$$

On the other hand, if food consumption is greater than basal metabolism ($C > \mu B$), the excess, $C - \mu B$, is used in active metabolism, a fraction of which is invested in production. Then

$$P = \varepsilon(C - \mu B) > 0, \tag{1.17}$$

where ε is the gross production efficiency for that part of consumption in excess of basal metabolism.

Since energy production is used for building new biomass, whether in the form of individual growth or reproduction, the contribution of the individual organism to the growth of biomass at the population level is P/γ, where γ is the energetic content of a unit biomass. Equation (1.17) can be scaled up to the population level if we make the simplifying assumption that all individuals are identical, as in classical population dynamical models. Subtracting losses due to mortality then yields

$$\frac{dB}{dt} = (\varepsilon/\gamma)C - (\varepsilon\mu/\gamma)B - \delta B, \tag{1.18}$$

where δ is the mass-specific death rate. In this equation I assume that $C > \mu B$, and hence production is positive, which holds as long as the population is not abruptly declining from starvation.

Consumption itself is a dynamical function of resource availability, which is described traditionally by the consumer functional response (Holling 1959). Let the consumer functional response to variations in the biomass of their resources, R, be defined here in the form of a mass-specific function $f(R)$. Further assume for simplicity that resources have the same energetic content γ as consumers. Energy consumption by the consumer population is then $\gamma f(R)B$. Substituting this expression into equation (1.18) yields

$$\frac{dB}{dt} = \varepsilon f(R)B - mB, \tag{1.19}$$

where $m = \varepsilon\mu/\gamma + \delta$ is a mass-specific loss rate which measures the long-term maintenance cost of a unit biomass, including both basal metabolism and mortality. This equation has the same form as that used in classical population dynamics to describe the dynamics of a consumer population. Therefore, it provides an explicit link between the functional approach used in ecosystem ecology and the demographic approach used in population and community ecology.

This simple equation provides a number of valuable insights. In particular, note that it has a structure similar to that of equation (1.4). Thus, it predicts exponential growth of the consumer population as long as its resources are abundant and roughly constant. Population regulation, however, is included indirectly in this equation through the consumer functional response since resource biomass R is a variable that decreases as the consumer population increases. There is no need to add density dependence in the form of a carrying capacity: density dependence arises spontaneously through the dynamics of the resources.

Equation (1.19) also yields insights into the functional meaning of traditional demographic parameters. Of special interest is parameter m, the mass-specific loss rate, which is often interpreted as a mortality rate. This rate, however, includes both death due to starvation (failure to meet basal metabolism) and natural death from other causes. Population or community ecologists sometimes assume implicitly that the per capita death rate in their models represents natural death, but this rate may differ by several orders of magnitude from the rate at which individuals die once they are deprived of food. For instance, in humans, the life expectancy of well-fed individuals is about 70 years, but that of starved individuals is only a few weeks—a difference of more than three orders of magnitude! This shows that a functional perspective is important to avoid misinterpreting demographic parameters.

Note also, for terminological clarity, that the coefficient ε in equation (1.19) has often been interpreted as the consumer's energy assimilation efficiency in the ecological literature (e.g., DeAngelis 1975; Yodzis and Innes 1992). The above derivation, however, shows that it actually represents its gross production efficiency for that part of consumption in excess of basal metabolism. I shall call it "production efficiency" in short in later chapters, although it does not correspond exactly to the definition of production efficiency in the energy budget literature.

As with any model, the great strength of a simple equation such as (1.19) is that it provides simple predictions and clear interpretations. Its corresponding weakness, of course, is that it does not provide a complete description of reality. Two important limitations need to be discussed here because they can have significant consequences for the dynamics of the model populations, communities, and ecosystems built on this equation.

First, equation (1.19) is based on the simplifying assumption that functional and demographic processes in the consumer population, such as consumption and mortality, are proportional to consumer biomass. Some authors (e.g., Owen-Smith 2002) have argued that the rate of death due to starvation should be a nonlinear function of food consumption because mortality increases steeply as food consumption decreases. In fact, a comparison of equations (1.16) and (1.17) shows that both the growth and loss terms in equation (1.19) should increase by roughly a factor $1/\varepsilon$ when most of the population starves. Although the assumption of a constant mass-specific loss rate is obviously a simplification—as is any other feature of equation (1.19) or of any other model—it is nevertheless a reasonable one at the population level as long as starvation is not acute, because the dependence of net population growth on resource availability is already captured in the consumption term. When resource availability R is insufficient for consumption to compensate for maintenance costs $[\varepsilon f(R) < m]$, the net population growth rate becomes negative, which amounts to an abrupt switching from growth to decline at the population level. A much stronger assumption is the lack of dependence of the mass-specific rates on population density or biomass in equation (1.19), which amounts to assuming that there is no interference among consumers, whether in the consumption or in the mortality process. There is no doubt that mutual interference does exist and can affect the dynamics of populations and communities qualitatively (DeAngelis et al. 1975; Arditi and Ginzburg 1989). Its prevalence and strength in nature, however, are controversial (Abrams and Ginzburg 2000), and its incorporation in population or ecosystem models complicates their analysis considerably. For the sake of simplicity, I shall accept in most of this book the traditional assumption that interference is negligible in trophic interactions.

Second, I have made explicit above another important assumption that is implicit in simple population dynamical models; i.e., all individuals are identical. This assumption is made for convenience because populations, or even whole functional groups (groups of species with similar functional roles in the ecosystem), will often be the basic unit in my representation of communities and ecosystems. This assumption is valid only to the extent that variation among individuals within a species or functional group is smaller than variation among species or functional groups. Alternative approaches when variation among individuals is significant include individual-based models (Huston et al. 1988; DeAngelis and Gross 1992) and physiologically structured population models (Metz and Diekmann 1986; De Roos et al. 2003). These approaches have greater realism and flexibility, but they are also more complex and parameter-rich. Accordingly, they are generally applied to more specific situations in which detailed information on individual behavior and ontogeny is available and plays an important role in population dynamics. I shall ignore individual variability within populations in the rest of this book.

Last, I have provided a functional derivation and interpretation of the demographic equation (1.19) above based on animal energy budgets. Most of the models that I shall present in this book, however, will involve plants as the basal living compartment of ecosystems and will be based on mass budgets tracking the fate of limiting nutrients. It is straightforward to generalize the above approach to these situations. Plant energy budgets are traditionally defined differently than animal energy budgets, but they comprise essentially the same elements. Evapotranspiration is the part of the absorbed energy that is not assimilated by plants and thus is the functional equivalent of egestion in animals. Gross primary production and net primary production in plants correspond to assimilation and production, respectively, in animal energy budgets. Excretion is often ignored in plant energy budgets; it is implicitly regarded as a loss to net primary production.

Mass budgets have been less studied than energy budgets. For elements other than carbon, the main difference is that there is no equivalent for respiration. Otherwise, similar derivations of population-level dynamical equations are possible in principle for nutrients just as for energy. Primary production is thought to be limited by nitrogen or phosphorus in most ecosystems. Since nitrogen and phosphorus are not directly involved in the chemical reactions of photosynthesis and respiration, I shall assume in the rest of this book that the uptake of limiting nutrients by plants is proportional to net primary production, which is the equivalent of production (plus excretion) in animal energy budgets since these nutrients are used for growth and reproduction. Mass and energy transfers are simultaneous in animals since consumed food contains both energy and materials. Therefore, equations similar to equation (1.19) can be derived for the dynamics of nutrient stocks in animal populations.

A MINIMAL ECOSYSTEM MODEL

The above population dynamical model based on mass or energy budgets suggests a simple way to build ecosystem models that satisfy both the physical laws of conservation of mass and energy and the demographic law of exponential growth at low population density or high resource availability: use the formalism of compartmental models but allow the dynamics of each compartment to be nonlinear functions of compartment sizes. All the complexity of biological interactions can be included in ecosystem models using this simple rule.

As the simplest possible application of this approach, consider a nutrient-limited ecosystem in which there is a single plant compartment with size P and an inorganic nutrient compartment with size N (H. T. Odum 1983). The size of each compartment is here measured by its nutrient stock. Assume that the ecosystem has a closed nutrient cycle (no input or output of nutrient) and that nutrient uptake by plants follows the law of mass action, i.e., is proportional to the product of P and N, as in standard Lotka–Volterra models (which corresponds to the linear part of a Holling type-1 functional response). The dynamics of the system can be written as

$$\frac{dP}{dt} = uNP - mP,$$
$$\frac{dN}{dt} = mP - uNP,$$

(1.20)

where u is the rate of nutrient uptake by plants per unit time per unit mass of nutrient, and m is the turnover rate of nutrient in plants due to basal metabolism and mortality.

Since the nutrient cycle is closed, any inflow to one compartment is an outflow from the other compartment, so that the equations for P and N are mirror images of each other. Summing the two equations, we see that the total quantity of nutrient in the system, $P + N$, is a constant, which I call Q:

$$\frac{d(P + N)}{dt} = 0,$$
$$P + N = Q.$$

(1.21)

This conservation equation can now be used to substitute $Q - P$ for N in the first of equations (1.20), yielding

$$\frac{dP}{dt} = rP\left(1 - \frac{P}{K}\right),$$

(1.22)

where $r = uK$, and $K = Q - m/u$.

This is nothing else than the familiar logistic equation of population dynamics. Thus, we see that the logistic equation can be obtained as the result of explicit nutrient limitation in a closed ecosystem. On a more technical note, notice how the mass conservation constraint in a closed ecosystem

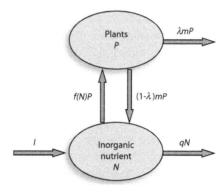

FIGURE 1.3. A minimal open, nutrient-limited ecosystem model. Circles represent nutrient stocks, while arrows represent nutrient flows.

reduces the effective dimensionality of the corresponding dynamical system (here, from two- to one-dimensional). Although incorporation of explicit ecosystem-level constraints may make population and community models look more complex at first sight, it may actually simplify their analysis under some conditions, just as incorporation of explicit resource dynamics may simplify the analysis of models of exploitation competition (Tilman 1982; Grover 1997). The reduction in dimensionality that results from incorporation of ecosystem-level mass-balance constraints is a trick that has been used in a number of theoretical studies in community and ecosystem ecology (e.g., Grover 1994; Holt et al. 1994; Loreau 1995).

The assumption of complete ecosystem closure to material exchanges with the outside world is of course unrealistic. A minimal ecosystem model that accounts for nutrient exchanges across ecosystem boundaries can be constructed as follows (figure 1.3). Assume that the inorganic nutrient pool is supplied with a constant input I of nutrient per unit time through processes such as water flow, dry deposition, and rock weathering and loses nutrient at a rate q per unit time through processes such as water flow, leaching, and volatilization. A fraction λ of nutrient is also lost from the ecosystem once released by plants, either before (e.g., through fire) or during (e.g., through leaching) the decomposition process. Let $f(N)$ denote the functional response of plants to nutrient availability and let m denote the rate at which they release nutrient because of basal metabolism and mortality as before. The resulting ecosystem model is a nonlinear version of the abstract two-compartment model depicted in figure 1.2.

The diagram depicting these processes (figure 1.3) translates into mathematical equations by applying the principle of mass conservation and setting the time derivative of compartment size equal to the sum of inflows minus the sum of outflows for each compartment. This provides the set of equations

$$\frac{dN}{dt} = I - qN - f(N)P + (1 - \lambda)mP,$$
$$\frac{dP}{dt} = f(N)P - mP. \tag{1.23}$$

In the long run, this dynamical system reaches an equilibrium. This occurs when inflows balance outflows for each compartment, and hence the time derivatives in equations (1.23) vanish. Solving the resulting mass-balance equations provides the equilibrium nutrient stocks (denoted by an asterisk as before)

$$N^* = f^{-1}(m),$$
$$P^* = \frac{I - qN^*}{\lambda m}, \tag{1.24}$$

where f^{-1} denotes the inverse function of f, the plant functional response. It is easy to check, using standard graphical or mathematical analyses, that this equilibrium is always stable (May 1973; Puccia and Levins 1985).

Although the dynamics of this system can no longer be reduced to a simple logistic equation for plants as with model (1.20), indirect density dependence of plant growth also occurs through nutrient limitation in this case, leading to regulation of plant nutrient stock (and hence biomass) around an equilibrium value or "carrying capacity." This carrying capacity [equation (1.24)] is now determined by the parameters that govern the plant–nutrient interaction and by the parameters that govern nutrient exchanges across ecosystem boundaries.

Model (1.23) also allows analysis and prediction of primary production. Since net primary production generally increases in proportion to plant nutrient uptake, the nutrient flow corresponding to plant nutrient uptake, $f(N)P$, can be used to measure net primary production, Φ_p. At equilibrium, the latter is simply

$$\Phi_p^* = mP^* = \frac{I - qN^*}{\lambda}. \tag{1.25}$$

This equation is easily interpreted. The numerator on the right-hand side of this equation is the excess of inflow of inorganic nutrient over its outflow at equilibrium; therefore it represents the net supply of nutrient in inorganic form available to plants at equilibrium. The denominator measures the fraction of nutrient lost from the plant compartment. Thus, equilibrium primary production is the product of two terms: (1) the net supply of the limiting nutrient, and (2) the efficiency with which this limiting nutrient is conserved by plants within the ecosystem (as measured by the inverse of λ). Equilibrium plant biomass [equation (1.24)] is then obtained simply by dividing primary production by the turnover rate of nutrient in plants. The implications of these equations will be further discussed in chapter 6.

CONCLUSION

Although the conceptual and formal foundations of population dynamics and ecosystem functioning are very different, both are related, directly or indirectly, to the mass and energy budgets of individual organisms. The dynamics of a species' biomass is determined by the way individuals allocate the nutrients and energy they consume to various physiological and behavioral processes, which allows the demographic parameters of population dynamics to be given a functional interpretation in terms of mass and energy flows. In turn, the flows of mass and energy in an ecosystem are determined by the population dynamics and interactions of its component species, which makes it possible to incorporate the complexity of demographic processes in the functions that govern mass and energy flows. Consistent models that merge the community and ecosystem perspectives can then be obtained by coupling the formalism of compartmental models borrowed from ecosystem ecology and the versatility of nonlinear functions that determine mass and energy flows borrowed from population and community ecology.

Armed with these principles and methods, we may now examine more thoroughly the processes involved in the organization of ecosystems, and the causes and consequences of community-level processes such as biodiversity changes and species interactions within ecosystems.

The Maintenance and Functional Consequences of Species Diversity

The core of community ecology is concerned with the question: why are there so many species on Earth? The tremendous diversity of life despite common constraints on the physiology and ecology of organisms is one of the hallmarks of living systems. Community ecology seeks to explain the maintenance of species diversity within ecological systems very much like population genetics seeks to explain the maintenance of genetic diversity within species. A large part of this diversity can be explained by geographical differences in environmental conditions across the globe and by historical circumstances. Many species and genetic variants, however, coexist in any given place and at any given time. Why do so many species and types coexist?

There are two main components to local species diversity, which I shall call vertical and horizontal, respectively. *Vertical diversity* is the diversity of functionally different types of organisms as defined by their trophic relationships or by other, nontrophic interactions (trophic levels, guilds, functional groups). The term "vertical" comes from the traditional representation of food chains in the form of vertical chains with plants at the bottom and carnivores at the top. By contrast, *horizontal diversity* is the diversity of species within trophic levels or functional groups. Vertical diversity concerns food webs and interaction networks and will be addressed in chapter 4. In this chapter I shall focus on the maintenance of horizontal diversity within ecological communities.

Explaining the coexistence of species with similar functional roles, or ecological niches, is the subject of *competition theory*. Competition theory is initially an extension of the theory of density dependence in population dynamics, in which intraspecific competitive interactions among individuals of a single species are extended to include interspecific competition among individuals of different species. This theory, however, has expanded considerably during the last 40 years to become a huge research field which has itself proliferated into a diversity of often competing theories.

My objective in this chapter is not to provide an exhaustive review of these theories, an enterprise that would be beyond the scope of this book. My objective is rather to examine their foundations in order to make sense of their commonalities and differences and understand their consequences for the functioning of communities and ecosystems. Competition theory is key to establishing a transition from populations to ecosystems because it deals with the first step in this transition; i.e., it links populations and aggregate community properties within functional groups. The second step in this transition is to link functional groups and overall ecosystem functioning, a step that will be made later in chapters 4 and 6. Therefore, although the present chapter does not deal with ecosystem functioning strictly speaking, the material it contains serves as a basis for subsequent chapters. In particular, it serves as a direct introduction to the next chapter, which examines the relationship between biodiversity and ecosystem functioning in simple systems with a single trophic level.

NICHE THEORY AND THE COEXISTENCE OF SPECIES IN HOMOGENEOUS ENVIRONMENTS

PHENOMENOLOGICAL MODELS

Phenomenological models are models that describe a phenomenon without explicit consideration of the mechanisms that generate this phenomenon. These models may be very apt at reproducing an observation, but they usually have limited predictive power outside the conditions that gave rise to this observation. The classical Lotka–Volterra competition model (Lotka 1925; Volterra 1926) is an example of this phenomenological approach. Although this model can be given a number of different mechanistic interpretations (some of which will be presented later in this chapter), it is not tailored to represent any specific mechanism a priori. In fact, it is a straightforward extension of the logistic equation to interspecific competition. Just like the logistic equation, it is based on a linear approximation to any function describing the dependence of the per capita population growth rate on the densities of the S species in competition:

$$\frac{dN_i}{dt} = \frac{r_i N_i}{K_i}\left(K_i - \sum_{j=1}^{S} \alpha_{ij} N_j\right). \tag{2.1}$$

In this equation, N_i is the population size of species i, r_i is its intrinsic rate of natural increase, K_i is its carrying capacity, and α_{ij} are competition coefficients. All variables and parameters have the same meaning as in the logistic equation, except for the competition coefficients. The latter are defined such that α_{ij} measures the competitive effect of species j on species i relative to the competitive effect of species i on itself. By this definition, the intraspecific coefficients $\alpha_{ii} = 1$, and the logistic equation is recovered when only intraspecific competition occurs.

The behavior of the Lotka–Volterra model is well known for two competing species and can be found in any ecology textbook. The two-species model can be analyzed using an isocline analysis, i.e., a graphical analysis that portrays the curves for which population growth is zero for the two species (known as null isoclines or zero net growth isoclines) in the phase or state plane (N_1, N_2). Each species has two null isoclines in this plane: $N_i = 0$, and $N_i = K_i - \alpha_{ij}N_j$, which are the two solutions of equation (2.1) when the population growth rate of species i is set to zero $(dN_i/dt = 0)$. It is easy to see from equation (2.1) that the population growth rate of species i is positive below the nontrivial isocline $N_i = K_i - \alpha_{ij}N_j$ and negative above it. This allows four types of asymptotic behaviors to be identified depending on the respective positions of the two nontrivial isoclines: (1) competitive exclusion of species 1 by species 2; (2) competitive exclusion of species 2 by species 1; (3) competitive exclusion of either species 1 or species 2 depending on initial conditions (with an unstable internal equilibrium point and two alternative stable states); and (4) stable coexistence of the two species. The configuration that leads to stable coexistence is shown in figure 2.1. In this configuration, the two nontrivial isoclines intersect in the positive quadrant, which ensures a feasible internal equilibrium point, and the isocline of species 1 has a steeper slope than that of species 2, which ensures convergence of the system's trajectories toward the internal equilibrium. For this configuration to occur, it is necessary and sufficient that the y-intercept of the isocline of species 1 be greater than that of species 2 $(K_1/\alpha_{12} > K_2)$ and that the reverse be true for the x-intercepts $(K_2/\alpha_{21} > K_1)$, which can be combined in the double inequality

$$\alpha_{21} < x < 1/\alpha_{12}, \qquad (2.2)$$

where $x = K_2/K_1$.

This condition for *stable coexistence* can be broken down into two distinct conditions:

1. *a feasibility condition* bearing on the ratio K_2/K_1, which ensures that the internal equilibrium exists (this ratio must be comprised in the interval $[\alpha_{21}, 1/\alpha_{12}]$);

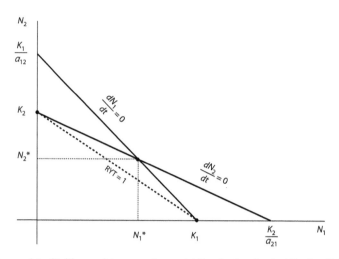

FIGURE 2.1. Stable coexistence and overyielding in the classical Lotka–Volterra competition model. Stable coexistence occurs when the null isoclines of the two species intersect as shown in the figure, which requires $K_2/\alpha_{21} > K_1$ and $K_1/\alpha_{12} > K_2$. The stable equilibrium point (N_1^*, N_2^*) then lies above the line RYT = 1 (dashed line) that connects the two monoculture equilibria $(K_1, 0)$ and $(0, K_2)$, which ensures overyielding. Reprinted from Loreau (2004).

2. *a stability condition* bearing on the competition coefficients (which determine the slopes of the isoclines), which ensures that the system converges to the internal equilibrium if it exists. This stability condition, which involves the two extreme terms in inequality (2.2), can be written as

$$\alpha_{12}\alpha_{21} < 1. \tag{2.3}$$

Since the intraspecific competition coefficients are set equal to 1 by definition, the stability condition states that interspecific competition (as measured by the geometric mean of the interspecific competition coefficients) must be smaller than intraspecific competition. Because of its intuitive biological interpretation, it is essentially this condition that attracted the attention of experimental, empirical, and theoretical ecologists during the development of competition theory in the 20th century. Gause (1934), in his classical experimental tests of the Lotka–Volterra model, had already supposed that the strength of interspecific competition was related to niche overlap, specifically overlap in resource use. This intuition allowed him to confirm the stable coexistence of competitors under the conditions predicted by the model. Combined with Volterra's (1926) mathematical proof, this led to formulation of the celebrated *competitive exclusion principle*

(Hardin 1960), which states that two species that occupy the same ecological niche cannot coexist indefinitely. As a matter of fact, Gause's experimental work did not provide strong evidence for this principle because mechanisms other than niche differentiation might have operated. But many other experiments have since confirmed it in constant, homogeneous environments under laboratory conditions (Arthur 1987).

The full condition for stable coexistence [inequality (2.2)], however, requires both interspecific competition to be smaller than intraspecific competition (stability condition) and the carrying capacities of the two species to be sufficiently similar to each other (feasibility condition). Thus, two types of mechanisms are involved in stable coexistence: equalizing and stabilizing mechanisms (Chesson 2000b). *Equalizing mechanisms* reduce the magnitude of the fitness difference between species (here determined by their carrying capacities), while *stabilizing mechanisms* concentrate intraspecific effects relative to interspecific effects. The two mechanisms must be present simultaneously to ensure stable coexistence.

As in any approach, the phenomenological approach, of which the Lotka–Volterra model is an archetype, has both strengths and weaknesses. Its main strength lies in its generality: since it ignores specific mechanisms, it may serve as an approximation to many different systems. In particular, the above stability condition carries over to the local stability of equilibrium points in more complex models.[1] The corresponding weakness of the Lotka–Volterra model is that it is not truly predictive. Parameter estimates are necessary to predict the outcome of interspecific competition between any two species. While intrinsic rates of natural increase and carrying capacities can be estimated from single-species measurements, competition coefficients cannot. Therefore, there is no way to predict a priori the outcome of a competition experiment; the model can only be fitted to the observations a posteriori.

Another, related weakness of the Lotka–Volterra model is that, just like the logistic equation of which it is an extension, it ignores mass and energy balance constraints and hence says nothing about functional processes such as consumption and production which would allow making explicit predictions on the consequences of interspecific competition for ecosystem functioning. The model, however, does make simple, powerful predictions regarding population size and hence biomass at the community level.

---

[1] This does not hold, however, for the feasibility condition since the feasibility of an equilibrium point hinges on the shape of the null isoclines, which in turn hinges on the detailed functional form of the dynamical equations.

For organisms such as annual plants in which yearly peak biomass is fairly well correlated with yearly production, this provides a valuable opportunity to explore ecosystem-level impacts of competition (Loreau 2004).

The theory of plant competition experiments developed in agricultural sciences provides a criterion to assess whether a mixture of two plant species shows *overyielding*, i.e., whether it yields more than expected based on their yields in monoculture (De Wit 1960; De Wit and van der Bergh 1965). If the two species use the same resource without niche differentiation, the increase in the yield of one species should be accompanied by a corresponding decrease in the yield of the other species, such that their relative yield total (RYT) is constant and equal to 1:

$$\text{RYT} = \frac{N_1}{K_1} + \frac{N_2}{K_2} = 1, \tag{2.4}$$

where N_1 and N_2 are the respective yields of species 1 and 2 in mixture, and K_1 and K_2 are their yields in monoculture.

When yield is simply measured by biomass, as is often done in annual plants, a graphical analysis shows easily that the equilibrium point corresponding to stable coexistence in the system lies above the straight line RYT = 1 that connects the two monoculture equilibria $(K_1, 0)$ and $(0, K_2)$ (Vandermeer 1989) (figure 2.1). Thus, stable coexistence necessarily implies RYT > 1, and hence overyielding, in the Lotka–Volterra model. Interspecific competition between species that have the potential to coexist owing to some form of niche differentiation leads to a community that yields more biomass than expected from the properties of its component populations. In other words, niche differentiation provides the basis for functional complementarity between species.

Overyielding, however, does not necessarily imply that the mixture outperforms the highest yielding monoculture, a phenomenon known as *transgressive overyielding*. Assume, without any loss of generality, that species 2 has the highest carrying capacity; i.e., $x \geq 1$ in inequality (2.2). Transgressive overyielding then occurs at equilibrium in the Lotka–Volterra model when

$$N_1^* + N_2^* > K_2 = xK_1. \tag{2.5}$$

The equilibrium values N_1^* and N_2^* are easily obtained by solving equations (2.1) after setting the time derivatives to zero. Substituting them into inequality (2.5) yields

$$(1 - \alpha_{21})(1 - x\alpha_{12}) > 0. \tag{2.6}$$

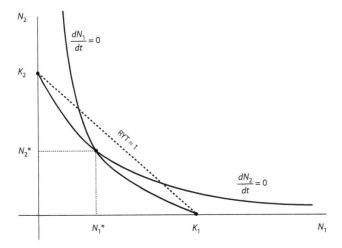

FIGURE 2.2. Stable coexistence in a model with convex nonlinear null isoclines (solid lines). In this case, the stable equilibrium point $(N_1{}^*, N_2{}^*)$ may lie below the line RYT = 1 (dashed line) that connects the two monoculture equilibria $(K_1, 0)$ and $(0, K_2)$. Reprinted from Loreau (2004).

Since the second term in parentheses on the left-hand side of (2.6) is positive by inequality (2.2), this condition reduces to

$$\alpha_{21} < 1. \tag{2.7}$$

It is straightforward to show that the coexistence condition (2.2) also necessarily implies $\alpha_{12} < 1$ and hence overyielding of the poorest yielding monoculture by the mixture. But it does not necessarily imply $\alpha_{21} < 1$ and hence overyielding of the highest yielding monoculture by the mixture. As equation (2.7) shows, transgressive overyielding further requires that interspecific competition be smaller than intraspecific competition in both species. This in turn requires stronger niche differentiation than is required for stable coexistence and nontransgressive overyielding since the latter are compatible with $\alpha_{21} > 1$.

Unfortunately, these conclusions obtained for the Lotka–Volterra model do not easily extend to more complex competition models. As mentioned above, the Lotka–Volterra model may be viewed as providing a linear approximation of per capita population growth rates, which may more generally be nonlinear. It is this linearity of per capita population growth rates that generates the linearity of the null isoclines (figure 2.1). Convex nonlinear isoclines allow stable equilibrium points to lie on or below the straight line RYT = 1 that connects the two monoculture equilibria (figure 2.2).

Under these conditions, stable coexistence may entail absence of overyielding, or even underyielding. Convex nonlinear isoclines were demonstrated in a *Drosophila* experimental system precisely to explain why stable coexistence is compatible with underyielding and why this does not invalidate competition theory (Gilpin and Justice 1972). How frequent they are in nature is still largely unknown.

Overyielding has been extensively used and much debated within the context of recent experiments on the functional consequences of biodiversity for ecosystem functioning, which will be considered in the next chapter.

MacArthur's Niche Theory

Overcoming the limitations of phenomenological models requires *mechanistic approaches*, i.e., approaches that explicitly consider the lower-level processes that generate the phenomenon considered (Schoener 1986). There are two main such approaches for interspecific competition, both of which concern exploitation competition, i.e., mutual negative effects among consumers that arise from exploitation of a joint array of resources. The first, which I shall examine in this section, is the niche theory developed by MacArthur and Levins (1967), Levins (1968), and MacArthur (1969, 1970, 1972); the second, which I shall consider in the following section, is the theory of limiting resources.

MacArthur's *niche theory* is a very elegant and powerful theory that applies to consumers that exploit and partition a set of substitutable resources along a resource gradient. The elegance and power of this theory comes from the fact that it addresses not only the composition of communities (what combinations of species win the competition) but also potentially their structure (niche structure, species abundance patterns) and functioning (energy flow). Since plants use nonsubstitutable, essential resources, this theory was developed mainly for animals. Indeed, most of the examples used by MacArthur concerned birds or lizards, which partition food along a body size gradient. In my brief presentation of the theory, I follow Gatto (1990), who generalized MacArthur's (1969, 1970, 1972) minimum principle for competitive communities.

The theory starts with an explicit representation of the dynamics of consumer–resource interactions. Let N_i be the abundance of consumer species i, and $R(z)$ the abundance of resources along a continuous gradient z, which measures, say, body size, or any other continuous trait of resources that allow consumers to discriminate among them. Assume that resources have logistic growth in the absence of consumers, resources are substitutable (such that their consumption rates by each consumer are additive), and

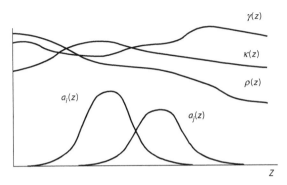

FIGURE 2.3. MacArthur's niche model based on explicit consumer–resource dynamics. The niches of consumer species i and j are described by their consumption functions $a_i(z)$ and $a_j(z)$ along the resource gradient z. The intrinsic rate of natural increase, $\rho(z)$, the carrying capacity, $\kappa(z)$, and the energy content, $\gamma(z)$, of resources may also vary along the gradient.

consumers have simple linear functional responses (corresponding to the linear part of a type-1 functional response, as in Lotka–Volterra models). The dynamics of consumer–resource interactions can then be written as

$$\frac{\partial R(z)}{\partial t} = \rho(z)R(z)\left[1 - \frac{R(z)}{\kappa(z)}\right] - \sum_{i=1}^{S} a_i(z)R(z)N_i, \tag{2.8a}$$

$$\frac{dN_i}{dt} = \eta_i \int \gamma(z)a_i(z)R(z)dz\,N_i - m_iN_i, \tag{2.8b}$$

where

$\rho(z) =$ intrinsic rate of natural increase of resource type z,
$\kappa(z) =$ carrying capacity of resource type z,
$a_i(z) =$ consumption rate of resource type z by a consumer of species i,
$\eta_i =$ coefficient of conversion of the energy consumed by consumer species i into new offspring,
$\gamma(z) =$ energy content of a unit of resource type z,
$m_i =$ mortality rate of consumer species i (due to starvation and natural death).

In this model, the niche of each consumer species i is described by its consumption function $a_i(z)$ along the resource gradient z (figure 2.3). But all the parameters that affect the growth and quality of resources, $\rho(z)$, $\kappa(z)$, and $\gamma(z)$, may also vary along the gradient (figure 2.3).

To reduce the dimensionality of this complex system, MacArthur made the additional assumption that resources have much faster dynamics than consumers, such that they are constantly at a moving equilibrium with consumers on the time scale of consumer dynamics. This amounts to setting $\partial R(z)/\partial t = 0$ in equation (2.8a). This equation can be solved for the equilibrium value of $R(z)$ as a function of N_i, which can be substituted into equation (2.8b) provided it is positive. MacArthur showed that the system then reduces to a classical Lotka–Volterra competition model for consumers as described by equation (2.1). But the parameters of this model now have a mechanistic interpretation in terms of consumer–resource interactions:

$$r_i = \eta_i \int \gamma(z)\kappa(z)\, a_i(z)\, dz - m_i, \qquad (2.9a)$$

$$K_i = \frac{\int \gamma(z)\kappa(z)a_i(z)dz - m_i/\eta_i}{\int w(z)a_i^2(z)dz}, \qquad (2.9b)$$

$$\alpha_{ij} = \frac{\int w(z)a_i(z)a_j(z)dz}{\int w(z)a_i^2(z)dz}, \qquad (2.9c)$$

where $w(z) = \gamma(z)\kappa(z)/\rho(z)$ is a weighting factor that measures the quality of resource type z, or, more specifically, its importance in exploitation competition among consumers: a resource that has a higher energy content, a higher carrying capacity, and a slower renewal (thus preventing fast replenishment of available resources) is expected to have a greater impact on intra- and interspecific competition among consumers.

A nice feature of MacArthur's model is that consumer demographic parameters have a simple energetic interpretation consistent with that presented in chapter 1 for single populations. Thus, r_i is consumer species i's intrinsic rate of natural increase, i.e., its per capita population growth rate when its abundance is negligible (close to zero) and hence resources are at their carrying capacity, $\kappa(z)$. In equation (2.9a), $\kappa(z)a_i(z)$ is the amount of resource type z consumed per consumer individual per unit time, which provides an amount of energy $\gamma(z)\kappa(z)a_i(z)$. Therefore, the integral in equation (2.9a) represents total per capita energy consumption per unit time. This energy consumption is converted to a birth rate through coefficient η_i. Last, subtracting the constant mortality rate m_i yields the per capita population growth rate of consumer species i when its abundance is negligible, as it should. The carrying capacity of consumer species i, K_i, is proportional to both (1) its per capita energy consumption in excess of its per capita energetic cost of population maintenance [the numerator in equation (2.9b)] and (2) a weighted measure of niche breadth. Indeed, the denominator in equation (2.9b) has a structure similar to the inverse of Levins's (1968) measure of niche breadth. The competition coefficient, α_{ij}, has a structure similar to Levins's (1968) measure of niche overlap. It can be interpreted as a ratio between the intensity of interspecific competition [measured by the numerator in equation (2.9c)] and the intensity of intraspecific competition [measured by the denominator in equation (2.9c)], in which resources are weighted by their quality as measured by $w(z)$.

A decisive advantage of MacArthur's mechanistic niche theory over the phenomenological Lotka–Volterra model is that it is *predictive*, not only descriptive, since all the parameters of the Lotka–Volterra model are in principle measurable a priori from empirical or experimental knowledge about the niche of each species and the properties of their resources. In particular, MacArthur (1972) analyzed the case where species have niches described by normalized Gaussian-shaped consumption curves along the resource gradient. Gaussian-shaped niches have been described in a number of bird and lizard species along a prey size gradient and should hold approximately in many instances after appropriate data transformation, given the general tendency for organisms to have an intermediate optimum in their response to environmental gradients. If σ_1 and σ_2 are the standard deviations of the consumption curves of species 1 and 2, respectively (a measure of niche breadth), and d is the distance between their means, the competition coefficients are then (MacArthur 1972; Case 2000).

$$\alpha_{ij} = \frac{\sqrt{2}\,\exp[d^2/2(\sigma_1^2 + \sigma_2^2)]}{\sigma_j\,\sqrt{(1/\sigma_1^2) + (1/\sigma_2^2)}} = \frac{\sigma_i\,\exp(d^2/4\overline{\sigma^2})}{\sqrt{\overline{\sigma^2}}}, \tag{2.10}$$

where $\overline{\sigma^2} = (\sigma_1^2 + \sigma_2^2)/2$. Thus, competition intensity drops steadily to zero as niche separation (as measured by the ratio $d/\sqrt{\overline{\sigma^2}}$) increases, and the potential for stable coexistence increases correspondingly.

As in any Lotka–Volterra system, stable coexistence entails overyielding. Transgressive overyielding, however, is not guaranteed. Here condition (2.7) becomes

$$\frac{d^2}{\overline{\sigma^2}} > 4\ln\left(\frac{\sigma_2}{\sqrt{\overline{\sigma^2}}}\right). \tag{2.11}$$

Thus, transgressive overyielding is facilitated by greater niche separation (larger ratio $d/\sqrt{\overline{\sigma^2}}$), and by greater similarity in niche breadth [when $\sigma_1 = \sigma_2, \sigma_2 = \sqrt{\overline{\sigma^2}}$, and the right-hand side of (2.11) vanishes]. When niches have identical breadth, any niche differentiation ($d > 0$) is sufficient to generate transgressive overyielding. On the other hand, when niches have identical optima ($d = 0$), no difference in niche breadth can make for transgressive overyielding.

But the theory is more powerful than this. A key finding of MacArthur is that a competitive community obeying equations (2.1) and (2.9) has a unique, globally stable equilibrium point, and the latter has biologically interpretable structural and functional properties. MacArthur (1969, 1970, 1972) and Gatto (1990) showed that such a community admits a Lyapunov function, i.e., a function that is always positive and whose time derivative is always negative, except at equilibrium where it is zero. Lyapunov functions are particularly useful because they ensure that the dynamical system converges to a unique equilibrium point where they are minimized; thus, they ensure global stability of the equilibrium.

Specifically, MacArthur's Lyapunov function has the form

$$Q = U + B,$$

$$U = \frac{1}{2} \int w(z) \left[\rho(z) - \sum_{i=1}^{s} a_i(z) N_i \right]^2 dz, \qquad (2.12)$$

$$B = \sum_{i=1}^{S} \frac{m_i}{\eta_i} N_i$$

and can be interpreted as follows (Gatto 1990). U is a weighted mean squared difference between the maximum productivity of resources as measured by their intrinsic rate of natural increase, $\rho(z)$, and their total consumption, $\sum_{i=1}^{S} a_i(z) N_i$. Thus, U measures unutilized productivity. Minimizing it amounts to performing a least-square fit of resource consumption to available resource production. Since m_i/η_i represents the per capita energetic cost of population maintenance (the amount of energy lost to basal metabolism and natural death per unit time by an average individual of species i), B is the maintenance energetic cost for the entire community. Both U and B may be viewed as different ways in which the community "wastes" available energy. Thus, Q is a measure of inefficient energy use, and its minimization is a *principle of maximum efficiency of energy use* in competitive communities.

This is a strong result. MacArthur (1972) showed, using hypothetical examples in the special case where B is constant, how this principle of maximum efficiency of energy use could be applied in principle to predict the composition and species relative abundances of competitive communities. Thus, competition makes not only structural (niche differentiation, species abundance patterns) but also functional (most efficient energy use) properties emerge at the community scale, which supports the view of a competitive community as a full-fledged self-organized system within the broader context of the ecosystem as a whole.

The bipartition of Q into its two components U and B, however, does not lead to simple predictions about the ecological properties that are maximized or minimized in a competitive community. The minimization of U may often conflict with the minimization of B: a small, species-poor community of highly efficient individuals will keep the maintenance energetic cost low but leave plenty of resource productivity unutilized, while an abundant, diversified community will fully use resource productivity but spend a lot of energy in maintenance (Gatto 1990). Thus, a stable competitive community may result from the trade-off between two conflicting constraints. As a consequence, several strategies might be possible depending on environmental conditions and species traits. In general, however, one should expect U to be relatively high and impose strong constraints on community assembly at relatively low levels of species diversity; therefore, species diversity and energy utilization efficiency should be promoted under these conditions. What happens at higher levels of diversity might be more variable, although evolution toward greater species diversity and greater specialization is often predicted in constant or predictable environments (Levins 1968; Gatto 1990).

Despite its elegance, MacArthur's theory has remained largely untested. The importance of interspecific competition and patterns of niche differentiation predicted by niche theory have been the subject of numerous experimental (Connell 1983; Schoener 1983a; Goldberg and Barton 1992; Gurevitch et al. 1992) and empirical (Strong et al. 1984; Crowley 1992; Gotelli and Graves 1996; Gotelli and McCabe 2002) tests in the field. Overall, these studies have provided mixed support for the theory: interspecific competition and unambiguous patterns of niche differentiation do occur under natural conditions, but they often concern a subset of dominant species or ecologically related species that belong to the same guild within broader communities. This is a fairly reasonable conclusion given that competition should not be expected to be the sole factor at work in natural ecosystems. In contrast, to my knowledge no tests have been attempted on functional or quantitative predictions derived from the mechanistic foundations of the theory.

MacArthur's theory is based on a number of simplifying assumptions, some of which are explicit or fairly straightforward but others are not. In particular, two implicit assumptions may have far-reaching implications. First, MacArthur's consumer–resource model incorporates intraspecific competition among resources through the logistic growth term in equation (2.8a), but it ignores interspecific competition between resources that have different positions along the gradient z. Unless the resource gradient is made up of extremely specialized, independent species, this assumption is neither logical nor realistic. Interspecific competition between resources can lead to counterintuitive indirect effects in the system, including indirect mutualism instead of exploitation competition among consumers (Levine 1976). Second, MacArthur's analysis of his model implicitly assumed that resources all reach a feasible (i.e., positive) equilibrium, the value of which could then be substituted into the dynamical equations of consumers. But strong resource depletion by efficient consumers may lead to extinction of part of the resource gradient. This violates the conditions under which MacArthur's analysis is valid and can deeply alter the relationship between niche overlap and competition intensity among consumers (Abrams 1998; Abrams et al. 2008). These two assumptions strongly limit the generality of MacArthur's theory.

THE THEORY OF LIMITING RESOURCES

A second mechanistic approach to exploitation competition is what I call here the *theory of limiting resources*. This approach was also pioneered by MacArthur and Levins (1964) and MacArthur (1972) and then expanded by Léon and Tumpson (1975) and Tilman (1980, 1982). Grover (1997) provides a comprehensive review of this theory and its more recent developments. In this approach, the dynamics of a small number of discrete resources is considered explicitly, which allows precise predictions to be made about the outcome of competition among the consumers of these resources. Since plants are usually limited by a small number of resources (such as nitrogen and phosphorus), this approach has been mainly applied to plant competition.

Significant insights into the properties of competitive systems can be gained from the simplest case where there is a single resource, R, for which an arbitrary number S of consumers, N_i, compete. Here, the size of compartments R and N_i could be measured equivalently by their abundance, density, biomass, or nutrient stock given appropriate parameterization. Since the theory has been mainly applied to plants competing for inorganic nutrients, I shall consider that R represents the available stock of an inorganic nutrient, and N_i represents the biomass of plant species i. The inorganic nutrient is assumed to be supplied at a constant rate I from an external source and to be lost at a rate q per unit mass, as in chapter 1. If consumers compete only through resource exploitation (no interference competition), the dynamics of the consumer–resource system can be expressed in the form

$$\frac{dR}{dt} = I - qR - \sum_{i=1}^{S} N_i f_i(R)/\varepsilon_i, \qquad (2.13a)$$

$$\frac{dN_i}{dt} = [f_i(R) - m_i]N_i, \qquad (2.13b)$$

where $f_i(R)$ is the numerical response of consumer species i to resource availability, and m_i is its density-independent mortality rate. As in chapter 1, the consumer's numerical response is assumed to be proportional to its functional response, $f_i(R)/\varepsilon_i$. Parameter ε_i represents species i's biomass production per unit resource, which incorporates both conversion of nutrient into biomass and production efficiency. The only constraint on the numerical response is that it should be a monotonic increasing function ($f_i' \geq 0$); i.e., consumer growth increases, or at least does not decrease, as resource availability increases, which is generally expected for a limiting resource.

It is straightforward to see from equation (2.13b) that each consumer species i tends to an equilibrium such that

$$R^* = f_i^{-1}(m_i), \tag{2.14}$$

where f_i^{-1} is the inverse of function f_i. Thus, each consumer species tends to control the resource at an equilibrium level that is entirely determined by its own traits (incorporated into f_i and m_i). Since the resource R can have only one equilibrium value, all consumer species except one are competitively excluded. The species that wins the competition is the species with the lowest equilibrium resource requirement (R^*). All other species have negative growth rates and are driven to extinction at that resource level because the latter is too low to meet their own resource requirement (Volterra 1926; Hsu et al. 1977; Armstrong and MacGehee 1980). Note that the same result also holds for MacArthur's niche model: if all consumers have identical niches—i.e., use the same array of resources—the species that has the lowest resource requirement displaces all the other species at equilibrium (Gatto 1990).

This simple result has important implications. First, it provides the basis for a more precise formulation of the competitive exclusion principle: when resources limit consumer growth, no more than one consumer can persist indefinitely on a single resource in a constant environment. This formulation can easily be extended to limiting factors other than resources (Levin 1970). Of course, this formulation leaves open the issue of what constitutes a distinct resource, which is far from trivial (Abrams 1988). Second, the above result further predicts a priori which species will persist. And third, it implies that *resource utilization is maximized* by the species that wins the competition, a result that echoes MacArthur's minimum principle. This corollary is intuitively obvious since the species that depresses resource abundance most can only do so by consuming more of the resource. A formal proof is easily obtained by noting that at equilibrium resource consumption is equal to net resource supply, i.e., $I - qR^*$, in equation (2.13a).

Thus, the species with the lowest R^* also has the highest net resource supply and the highest resource consumption.

Note that the best competitor does not necessarily maximize other functional properties, such as production and biomass. Consumer production, which is given by $f_i(R)N_i$ in equation (2.13b), is equal to $\varepsilon_i(I - qR^*)$ at equilibrium from equation (2.13a); this is maximized when R^* is minimized only if all species have equal production coefficients ε_i. Equilibrium biomass, N_i^*, is equal to $(\varepsilon_i/m_i)(I - qR^*)$ from equations (2.13); this is maximized when R^* is minimized only if all species have equal production coefficients ε_i and mortality rates m_i that are either equal or vary parallel to R^*. If a single parameter varies among species, then competition should also generally maximize production and biomass. If, however, several parameters vary simultaneously among species, there is no guarantee that production and biomass will be maximized. As a matter of fact, simultaneous variation in several functional or demographic parameters is likely given the widespread occurrence of trade-offs between different physiological functions and life-history traits in ecology. And when resources are self-reproducing living organisms, the potential for nonlinear relationships between consumer production or biomass and equilibrium resource availability is even greater, as I shall show in chapter 4.

Extending the theory to two or more resources brings additional complexity for two main reasons. First, the consumer's numerical response to two resources can take on diverse forms, which determine different resource types (Léon and Tumpson 1975; Tilman 1980, 1982). At one extreme are perfectly substitutable resources, which is often assumed to be the case in animals faced with food items of roughly equal quality. At the other extreme are essential resources such as different chemical elements for plants, which cannot be substituted for one another. Different resource types generate different shapes of consumer isoclines in a phase plane determined by the two resources. Second, competition between two species with similar resource types leads to four types of outcomes very much as in the Lotka–Volterra model: competitive exclusion of one species, competitive exclusion of the other species, competitive exclusion of either species depending on initial conditions, and stable coexistence.

Let the two (plant) consumer species be A and B with biomasses N_A and N_B and let the two (inorganic) resources be 1 and 2 with stocks R_1 and R_2. Assume that resources obey the same dynamical equation as above; i.e.,

$$\frac{dR_i}{dt} = q(S_i - R_i) - \sum_{j=A}^{B} N_j\, c_{ji}. \tag{2.15}$$

Here, $S_i = I/q$ is the equilibrium amount that resource i would reach in the absence of consumption, and c_{ji} is the consumption rate of resource i by an individual consumer j, i.e., the functional response of consumer j. At equilibrium, net resource supply and resource consumption must balance, which can be written in vector form as

$$S = C_A + C_B, \tag{2.16}$$

where

$$S = q \begin{pmatrix} S_1 - R_1^* \\ S_2 - R_2^* \end{pmatrix}, \qquad C_A = N_A^* \begin{pmatrix} c_{A1}^* \\ c_{A2}^* \end{pmatrix}, \qquad C_B = N_B^* \begin{pmatrix} c_{B1}^* \\ c_{B2}^* \end{pmatrix}.$$

The configuration that leads to stable coexistence is shown in figure 2.4 for plants that use essential resources. When resources are essential, Liebig's law of the minimum holds; i.e., plant growth is limited by the nutrient that is in shortest supply irrespective of the abundance of the other nutrient, which results in L-shaped consumer isoclines in the (R_1, R_2) phase plane. Léon and Tumpson (1975) and Tilman (1980, 1982) showed that stable coexistence requires a number of conditions, specifically: (1) the consumer null isoclines must intersect in the positive quadrant, which requires that each species be the better competitor (lower R^*) for one of the resources; (2) the consumption vector, C_A, of species A, which is the better competitor for resource 1, must be steeper than the consumption vector, C_B, of species B, which requires that each species consume proportionately more of the resource that limits its own growth more; (3) the resource supply point S with coordinates (S_1, S_2) must fall in the region comprised between the two consumption vectors (figure 2.4).

When stable coexistence occurs, the resulting community makes better use of the two resources than would either species alone. At the coexistence equilibrium, species A sets R_2^* while species B sets R_1^* (figure 2.4); i.e., each resource is controlled by the inferior competitor for that resource. If species A were alone with a resource supply point in the region comprised between the two consumption vectors, the equilibrium point would shift to the right along the isocline of species A, which means that R_2^* would be unchanged while R_1^* would be higher. Similarly, if species B were alone, R_1^* would be unchanged while R_2^* would be higher. Thus, the two species collectively bring the two resources to lower levels than would occur if only one species were present. Equation (2.16) then shows that net resource supply and hence total resource consumption are higher in mixture than in monoculture. If the two species have equal production coefficients ε_i, this increased total resource consumption should also lead to an increased primary production of the mixed community. Thus, here again, we see that

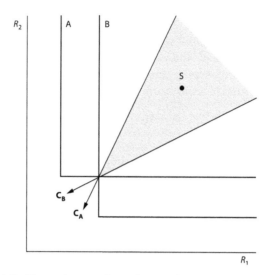

FIGURE 2.4. Stable coexistence of two plant species on two essential resources. Stable coexistence requires that (1) the plant null isoclines of the two species (indicated by A and B) intersect in the positive quadrant, i.e., each species be the better competitor (lower R^*) for one of the resources; (2) the consumption vector, $\mathbf{C_A}$, of species A, which is the better competitor for resource 1, be steeper than the consumption vector, $\mathbf{C_B}$, of species B; (3) the resource supply point S fall in the gray region comprised between the two consumption vectors. Under these conditions, the community makes better use of the two resources than would either species alone.

the conditions that promote stable coexistence also promote a better collective resource exploitation, which should generally result in increased total production by the community as a whole.

The theory of limiting resources can easily be extended to a broader theory of limiting factors, including predation, and provides the basis for a more comprehensive formal definition of the niche than in MacArthur's theory. It makes explicit the two components of a species' niche: its response to, or requirement from, the environment (here encapsulated in its null isocline), and its effect, or impact, on the environment (here encapsulated in its consumption vector) (Leibold 1995; Chesson 2000b; Chase and Leibold 2003).

Tilman (1982) provided experimental evidence that supports the predictions of the theory regarding the conditions of coexistence. Most of the experiments, however, have used freshwater algae under laboratory conditions. Experimental evidence for the theory under field conditions and for other organisms is still scarce.

COEXISTENCE IN SPATIALLY STRUCTURED ENVIRONMENTS

Classical theories in ecology were built on the conceptual and mathematical tools available from classical mechanics and chemistry to describe homogeneous systems (in particular, ordinary differential equations) and accordingly ignored spatial and temporal variations of ecological systems. By focusing on the behavior of idealized average systems, the implicit hope behind this approach was that it would uncover fundamental laws uncluttered by the noise generated by the variability of real systems. This analytical approach has been very successful in ecology, just as in physics or chemistry. In a number of cases, the laws or generalizations obtained for homogeneous systems still hold approximately for heterogeneous systems after appropriate adaptation of the formulation and interpretation of these laws or generalizations. However, it is increasingly clear in all sciences that variability is not simply noise. *Variability* per se, just like diversity, can add new dimensions, new constraints, and new opportunities, as has been amply demonstrated in ecology.

The study of temporal and especially spatial variations of ecological systems has grown tremendously during the last decades (Tilman and Kareiva 1997; Dieckmann et al. 2000). This book is not the appropriate place to review all these developments, which are too numerous and diversified. In this and the next section, I simply attempt to extract some of their main consequences for the maintenance and functional consequences of species diversity in competitive communities. I first focus on spatial variability in this section and then consider temporal variability in the next section.

The importance of *spatial structure* for coexistence has been recognized at least since Park's (1948) experiments on flour beetles, but it is only recently that formal quantitative theories have been developed to examine the mechanisms and outcomes of competitive coexistence in spatially structured environments (reviewed in Amarasekare 2003). These theories can be organized conceptually along two main axes.

First, they differ according to the spatial scale they consider: some theories focus on *regional processes*, while others consider only *local processes* to explain local coexistence. By "local" I mean here the scale that defines the system under consideration (whether a population, a community, or an ecosystem); in contrast, "regional" denotes entities or processes at larger spatial scales. Note that many theories can be applied equivalently to several spatial scales. Early models of competition in spatially structured environments (Levins and Culver 1971; Horn and MacArthur 1972; Slatkin 1974; Hanski 1981, 1983) considered patchy environments but were not explicit about scales or mechanisms. They were actually derived from models of metapopulations, i.e., of regional sets of local populations that are spatially distinct but connected by dispersal. Therefore, they could obviously be applied to community dynamics at landscape or regional scales. The difference between the theories that invoke regional processes and those that invoke local processes to explain local coexistence lies in the fact that the former consider cross-scale interactions explicitly: the processes that explain coexistence at the local scale are to be found at the larger regional scale rather than within the local system itself. Regional influences will be considered in chapter 7; in this section I shall consider only those theories that invoke processes at the scale where coexistence occurs.

Second, spatial structure can arise in two fundamentally different ways in spatial theories of coexistence: either *the environment is spatially heterogeneous* or the environment is homogeneous, and *spatial structure is created by the organisms themselves*. These two types of situations impose different constraints on coexistence: if spatial heterogeneity is a preexisting feature of the environment, organisms can respond to variations in environmental conditions in accordance with niche theory unless dispersal or recruitment limitation prevents them from doing so; if, on the contrary, the environment is homogeneous at the outset, some mechanism generating limited access of organisms to available space must be present to create and maintain spatial structure.

Before considering these two cases, it is useful to start with a background theory against which they can be contrasted (Loreau and Mouquet 1999). Suppose a homogeneous environment and a community of sessile organisms such as plants in which space occupancy obeys a competitive lottery; i.e., each individual occupies a distinct site and keeps it until it dies a natural death, and vacant sites are occupied by new individuals of each species in proportion to its contribution to a common pool of propagules. Let p_i be the proportion of sites occupied by species i. There are S such species that compete for a limited proportion of vacant sites, V. Each species i is characterized by a potential recruitment rate, c_i, which incorporates seed production, short-distance dispersal, germination and seedling establishment, and a mortality rate, m_i, which encapsulates all forms of natural death. Potential recruitment, however, is not fully realized because only vacant sites can be occupied. The dynamics of such a system is described by the equation

$$\frac{dp_i}{dt} = c_i p_i V - m_i p_i = m_i p_i (R_i V - 1), \qquad (2.17)$$

where $V = 1 - \sum_{j=1}^{S} p_j$, and $R_i = c_i/m_i$. R_i is known as species i's basic reproductive rate, but it is more aptly called its basic reproductive capacity since it is a dimensionless number. The basic reproductive capacity measures the average number of successful offspring that an individual produces during its lifetime in a vacant environment.

It is straightforward to see that each species i tends to an equilibrium such that

$$V^* = 1/R_i \qquad (2.18)$$

Thus, V^* plays the same role as R^* in the theory of limiting resources: the species with the lowest V^*, and hence with the highest basic reproductive capacity, displaces all the other species. In the absence of any other mechanism, spatial structure per se does not suffice to create the conditions for its own maintenance and hence for species coexistence. A competitive lottery leads to competitive exclusion because space is the single limiting resource in such a system.

Spatial Coexistence in Heterogeneous Environments

It is a relatively easy step, at least conceptually, to incorporate spatial heterogeneity into classical niche theory. If different species are adapted to different environmental conditions and these conditions vary within a locality or region, one would expect each species to dominate in those places where it is best adapted to the local environment, and hence species diversity to be maintained in the heterogeneous system as a whole. An example of such a mechanism is Tilman's (1982) resource ratio hypothesis. The theory of limiting resources shows that no more than two competitors can coexist at equilibrium on two resources under given environmental conditions, i.e., at a given resource supply ratio. Different species or combinations of species, however, may persist at different resource supply ratios. In this case, spatial variation in resource supply leads to global coexistence of species that would exclude each other at each location. Dispersal limitation interacts with competition, predation, and other species interactions to limit species distributions along environmental gradients and enhance regional species richness (Case et al. 2005).

Chesson (2000a) attempted to formalize a general model of competitive coexistence in spatially heterogeneous environments. Using this model, he showed that the two main spatial mechanisms of coexistence are analogous to corresponding temporal mechanisms of coexistence, which he identified as relative nonlinearity of competition and the storage effect (see below). These spatial mechanisms, however, involve different life-history traits, and suggest that the spatial storage effect should arise more commonly than the temporal storage effect, while spatial relative nonlinearity should arise less commonly than its temporal counterpart.

Coexistence in spatially heterogeneous environments should allow the species best adapted to particular environmental conditions to make the best use of the available resources under these conditions, which should lead to greater collective resource utilization efficiency by the community as a whole, just as in MacArthur's niche theory. Chesson et al. (2001) provided theoretical support for this intuitive idea and suggested that the enhanced performance of diverse plant communities in spatially heterogeneous environments also leads to increases in other ecosystem processes such as net primary production, carbon storage, nitrogen mineralization, and evapotranspiration.

SPATIAL COEXISTENCE IN HOMOGENEOUS ENVIRONMENTS

When no spatial heterogeneity preexists in the environment, the maintenance of spatial structure and species coexistence requires the operation of some mechanism that limits access of organisms to available space. The most common such mechanism is *recruitment or dispersal limitation*, which tends to generate clusters of conspecifics, thereby increasing the intensity of intraspecific competition relative to interspecific competition and thus the potential for coexistence. Coupled with spatial heterogeneity and niche differences among species, recruitment limitation is able to maintain high levels of species diversity (Hurtt and Pacala 1995). In the absence of spatial heterogeneity, at least two factors can operate in conjunction with recruitment limitation to allow indefinite coexistence of competitors: (1) life-history trade-offs (Bolker and Pacala 1999), and (2) differences in the spatial scales over which intra- and interspecific competition occur (Murrell and Law 2003).

The most familiar life-history trade-off allowing coexistence in a spatial context is the *competition–colonization trade-off*, from pioneer species that are good colonizers but poor competitors to climax species that are poor colonizers but good competitors (Levins and Culver 1971; Horn and Mac-Arthur 1972). Hastings (1980) and Tilman (1994) formalized it into a simple, spatially implicit model which is derived from classical metapopulation models but in which a single individual occupies a site as in the above competitive lottery. The difference with the competitive lottery is that here a superior competitor is assumed to displace an inferior competitor instantaneously if it reaches a site occupied by the latter. Thus, in addition to global competition for space, there is another form of competition, whether by exploitation or interference, that gives a strong local advantage to some species over others. Assuming a strict hierarchy in local competitive abilities such that species 1 is the best local competitor and species n is the poorest local competitor, this yields the following model:

$$\frac{dp_i}{dt} = c_i p_i \left(1 - \sum_{j=1}^{i} p_j\right) - m_i p_i - \sum_{j=1}^{i-1} c_j p_j p_i. \tag{2.19}$$

There are two differences between this equation and equation (2.17) for the competitive lottery. First, the amount of space perceived as vacant is different for each species: species i "sees" as occupied only those sites that are occupied by it or by better competitors, hence the summation in the first term on the right in equation (2.19) runs from species 1 to i. Sites occupied by inferior competitors (species $i + 1$ to S) count as vacant sites. Second, the last term on the right in equation (2.19) accounts for competitive displacement of species i by better competitors (species 1 to $i - 1$) as these reach sites occupied by species i.

Since the best competitor is unaffected by the other species, equation (2.19) reduces to Levins's (1969) metapopulation model for species 1. Therefore, its equilibrium space occupation is simply

$$p_1^* = 1 - \frac{m_1}{c_1}. \tag{2.20}$$

Species 2 is affected only by species 1 and by itself. Therefore, its equilibrium space occupation is easily obtained knowing that of species 1:

$$p_2^* = \frac{m_1}{c_1} - \frac{c_1 + m_2 - m_1}{c_2}. \tag{2.21}$$

The equilibrium space occupation of species 3 can then be obtained knowing those of the first two species, and so on for the other species. The constraints on species coexistence, however, can be understood qualitatively by considering only the first two species. The equilibrium space occupation of species 1 [equation (2.20)] is equal to the difference between the total proportion of sites available (1) and the proportion of sites that it leaves vacant at equilibrium (m_1/c_1). The equilibrium space occupation of species 2 in turn is equal to the proportion of sites left vacant by species 1 [the first term on the right-hand side of equation (2.21)] minus the proportion of sites that it leaves vacant at equilibrium [the second term on the right-hand side of equation (2.21)]. The proportion of sites left vacant by species 2 then becomes space available for species 3, and so on potentially ad infinitum. Since no species is ever able to occupy space fully, unlimited coexistence is possible in principle.

There are, however, strong constraints on this coexistence because each additional species must be able to occupy space better than the species that precedes it in the competitive hierarchy. It can do so in two ways: either by being better at recruiting at new sites (higher c) or by being better at keeping them in the absence of interspecific interactions (lower m). The first situation is more common because trade-offs between competitive ability and dispersal ability are thought to be widespread—hence the usual interpretation of this model as requiring a competition–colonization trade-off (Tilman 1994). Note that the term "colonization" is not strictly appropriate here since occupancy of a new site corresponds to the recruitment of a new individual in the population, not the establishment of a new population in a metapopulation as in the original formulation (Levins and Culver 1971).

More fundamentally, spatial structure per se is not what maintains diversity in this system. The competition–colonization trade-off is, in fact, a trade-off between two forms of competition: global competition for vacant space (in which the best "colonizers" win) and local competition by either interference or exploitation for another resource that is not represented in the model (in which the best "competitors" win). The model can then be reinterpreted in a nonspatial context as a model in which there is a trade-off between the ability for interference competition and the ability to use the shared resource (here, space) and hence maintain a higher carrying capacity (Adler and Mosquera 2000; Kokkoris et al. 2002). It is, in fact, formally identical to the classical Lotka–Volterra model with the following transformations (Loreau 2004):

$$
\begin{aligned}
r_i &= c_i - m_i, \\
K_i &= 1 - m_i/c_i, \\
\alpha_{ij} &= 1 + c_j/c_i, \quad j < i, \\
&= 0, \quad j > i.
\end{aligned}
\tag{2.22}
$$

This transformation helps in exploring some of the functional consequences of coexistence in a system maintained by the competition–colonization trade-off. In particular, when two species coexist by this mechanism, nontransgressive overyielding is ensured as in any Lotka–Volterra system with stable coexistence. In contrast, equations (2.22) show that the competition coefficient measuring interference from the species with the smaller carrying capacity (species 1) with the species with the larger carrying capacity (species 2) is greater than 1 as a result of the strong competitive asymmetry assumed in the model. This violates condition (2.7) and thus precludes transgressive overyielding.

Note, however, that yield is measured here by the fraction of space occupied by each species, which may have little to do with actual biomass production. It is likely that species that have a higher potential recruitment rate also have a higher productivity. Therefore, to be slightly more realistic while keeping simplicity, assume now that a species' production is given by its recruitment potential, which is the product of the number of sites it occupies and its potential recruitment rate at a site. Total production is then approximated by (Loreau and Mouquet 1999)

$$\Phi = \sum_{i=1}^{S} c_i p_i. \tag{2.23}$$

Since species 2 has both a higher spatial carrying capacity and a higher recruitment rate, it also has a higher product of these two quantities and hence a higher production. As a result, total production in the two-species community is intermediate between that of the two species when they are alone, and transgressive overyielding is again impossible with this measure of yield. On the other hand, nontransgressive overyielding is guaranteed since species 1 is unaffected by species 2, and hence the relative yield total is necessarily greater than 1 when species 2 coexists with species 1. Thus, a community structured by the competition–colonization trade-off has the peculiar properties that nontransgressive overyielding always occurs but transgressive overyielding never occurs.

Bolker and Pacala (1999) identified three main strategies in spatially homogeneous environments, i.e., colonization, exploitation, and tolerance, which are analogous to Grime's (1979) empirically based plant strategies. Life-history trade-offs other than the familiar competition (tolerance)–colonization trade-off are possible among these strategies. Although their functional consequences have not been explored, they are likely to be often similar. Life-history trade-offs may be viewed as a form of niche differentiation among species since they involve deterministic differences in the way these species interact with their environment. These niche differences tend to generate complementarity among species and hence overyielding. If the physical environment is homogeneous, however, space is a fixed resource to be shared among all species; its monopolization by superior exploiters is prevented by interactions with species that have some other competitive advantage such as interference competition. Transgressive overyielding, i.e., a community that uses the shared spatial resource better than does the best of its members, is unlikely under these conditions. In contrast, it is likely in the coexistence mechanism suggested by Murrell and Law (2003), i.e., heteromyopia. In this mechanism, the distance over which individuals see (i.e., interact with) their heterospecific neighbors is shorter than the distance over which they see their conspecific neighbors. This effectively creates small gaps in a landscape occupied by one species, in which another

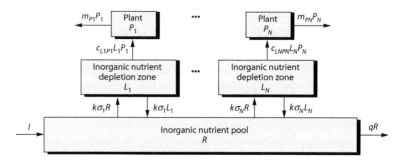

FIGURE 2.5. A simplified ecosystem model in which N plants have limited access to an inorganic nutrient in individual resource depletion zones. The shared inorganic nutrient pool is affected only indirectly by plants through diffusion and other nutrient transport processes.

species can spread. In this case, two species may better exploit space that any one of them separately.

A different mechanism of species coexistence in spatially homogeneous environments was suggested by Huston and DeAngelis (1994) and further analyzed by Loreau (1996, 1998a) and Goreaud et al. (2002). Individual organisms have only limited access to available resources because of physical limitations in body size and mobility. In particular, plants are sessile organisms that affect soil nutrients only in the immediate vicinity of their rooting system. Huston and DeAngelis (1994) proposed to model such a *resource access limitation* by distinguishing two types of resource compartments: an individual resource depletion zone under the direct control of each individual plant, and a global nutrient pool in the rest of the soil that is affected only indirectly by plants through diffusion and other nutrient transport processes. Variation in the rate of nutrient transport in the soil determines the degree to which plants interact competitively through their indirect effect on the shared nutrient pool.

To understand the main factors that control competitive interactions and species diversity in such a system, it is useful to consider a simplified model in which plants compete for a single limiting nutrient, and nutrient recycling and higher trophic levels are ignored (Loreau 1998a). A global soil inorganic nutrient pool (with concentration R and volume V_R) supports N individual plants i (with concentration P_i expressed per unit volume of their resource depletion zone), which exploit the soil inorganic nutrient in their individual resource depletion zones (with concentration L_i and volume V_i) (figure 2.5). Nutrient in inorganic form flows through the ecosystem at a rate q per unit time; R_0 is the inflowing nutrient concentration, defined as $R_0 = I/q$, where

I is the input of inorganic nutrient. Nutrient is transported by physical processes between the individual and global pools at a rate k per unit time. The global soil inorganic nutrient pool, R, then obeys the dynamical equation

$$\frac{dR}{dt} = q(R_0 - R) - k \sum_{i=1}^{N} \sigma_i (R - L_i), \tag{2.24}$$

where $\sigma_i = V_i / V_R$ is the relative volume of plant i's resource depletion zone.

For simplicity's sake, assume now that the resource depletion zones of all plants have identical volumes, $V_i = V$, and hence $\sigma_i = \sigma$. At equilibrium, the time derivative in equation (2.24) vanishes, which provides the following equilibrium concentration for the global soil inorganic nutrient pool:

$$R^* = \frac{qR_0 + k\sigma \sum_{i=1}^{N} L_i^*}{q + k\sigma N} = pR_0 + (1 - p)\overline{L^*}, \tag{2.25}$$

where $p = q/(q + k\sigma N)$, and $\overline{L^*}$ is the average nutrient concentration in the resource depletion zones across all plant individuals. Thus, the equilibrium nutrient concentration in the soil inorganic pool is a weighted average of R_0, its equilibrium concentration in the absence of plants, and $\overline{L^*}$, the average nutrient concentration of the resource depletion zones under plant control (with $\overline{L^*} < R^* < R_0$). The weighting coefficient p depends on the relative importance of the processes that govern global inputs and outputs in the system (q) compared with those that govern nutrient transport in the soil, i.e., the nutrient transport rate (k), and the number (N) and size (σ) of the plant resource depletion zones. If nutrient transport is very small (k tends to zero), plants have virtually no influence on the global soil nutrient concentration; as a result, competition is negligible and coexistence is unlimited. In contrast, if nutrient transport is very large (k tends to infinity), plants strongly affect the global soil nutrient concentration; in this case, competition among plants mediated by the global pool is intense, and the plant species with the lowest L^* can be shown to eventually outcompete all other plants just as in a nonspatial environment.

The strength of this individual-based approach lies in its ability to link individual- and ecosystem-level functional properties (chapter 3). Although it has only been applied so far to plants competing for soil nutrients, its area of application is potentially much broader. For instance, different kinds of insect herbivores, such as leaf eaters, stem borers and root feeders, have access to different plant parts and may coexist as a result of this form of niche differentiation. A plant could then be represented as a coupled system of plant parts similar to the system of individual and global nutrient pools discussed above. Likewise, predators may attack only part of the prey population if some prey have access to refuges. The prey population could again be divided into an accessible compartment and a nonaccessible compartment.

An important limitation of this approach, however, is that spatial structure is assumed a priori. While some forms of spatial structure may be dictated by inherited niche differentiation, this is generally not the case with plants competing for space. Thus, while this theory accounts for short-term plant coexistence, it does not explain long-term coexistence because it does not consider the spatial dynamics of site occupation. As a matter of fact, what maintained long-term plant coexistence in Huston and DeAngelis's (1994) simulations was the fact that they sampled new occupants of vacant sites from a regional species pool, thus mimicking an immigration process from a regional source (chapter 7).

NONEQUILIBRIUM COEXISTENCE IN TEMPORALLY FLUCTUATING ENVIRONMENTS

The problem of *nonequilibrium coexistence* was first introduced by Hutchinson (1961), who felt that niche theory, which he largely contributed to develop, was unable to explain what he called the "paradox of the plankton," i.e., the coexistence of a large number of algal species on a few limiting resources. Hutchinson reasoned that if the characteristic time scales of environmental variability and competitive exclusion are very different, competitive exclusion should proceed as in a constant environment because organisms either grow so fast that the best competitor displaces the other species before an environmental change takes place, or grow so slowly that they experience an average environment. On the other hand, if the characteristic time scales of environmental variability and competitive exclusion are similar, nonequilibrium coexistence of competitors on the same resources should be possible because environmental changes may provide a temporary competitive advantage to different species at different times.

While correct, this intuitive argument does not ensure that coexistence will be maintained in the long term since shifting competitive advantages may still lead to the progressive growth or decline of species over longer time scales. There are two main approaches to nonequilibrium coexistence with different implications, depending on whether coexistence is permanent or transient.

PERMANENT NONEQUILIBRIUM COEXISTENCE

Permanent nonequilibrium coexistence occurs when species fluctuate through time but persist indefinitely. Probably the most elegant, if nonintuitive, demonstration that variability per se is a factor that promotes permanent coexistence among species was provided by Levins (1979). Levins illustrated this by a simple model of two consumers, with population sizes N_1 and N_2, competing for a single resource with abundance R:

$$\frac{dN_1}{N_1 dt} = \frac{d \ln N_1}{dt} = R - m_1,$$

$$\frac{dN_2}{N_2 dt} = \frac{d \ln N_2}{dt} = R + R^2 - m_2. \tag{2.26}$$

Species 1's per capita population growth rate is a linear function of resource abundance as in classical Lotka–Volterra models, while species 2's is nonlinear. The quadratic term in R in species 2's per capita population growth rate has no biological justification; it is chosen because it is the simplest form of nonlinearity, and it is this nonlinearity that allows coexistence. Resource abundance is assumed to fluctuate through time because of either intrinsic factors (unstable consumer–resource interactions) or extrinsic factors (environmental forcing).

If the two species persist indefinitely at finite positive population sizes, the logarithm of population size is bounded away from $-\infty$ and $+\infty$, which ensures that the long-term time-averaged value of their time derivative is zero. In other words, if populations neither go extinct nor explode to infinite sizes, their abundance (and its logarithm) has to fluctuate around an average value such that its long-term average rate of change is zero. Denoting the temporal mean or expected value of a variable x by $E(x)$, this means that

$$E(R - m_1) = E(R) - m_1 = 0,$$

$$E(R + R^2 - m_2) = E(R) + E^2(R) + \text{var}(R) - m_2 = 0, \tag{2.27}$$

where the following property of statistical moments has been used:

$$\text{var}(R) = E(R^2) - E^2(R). \tag{2.28}$$

These equations have the solution

$$E(R) = m_1,$$

$$\text{var}(R) = m_2 - m_1(m_1 + 1), \tag{2.29}$$

which is feasible provided that $m_2 > m_1(m_1 + 1)$, or, equivalently,

$$\sqrt{m_2 - m_1} > m_1. \tag{2.30}$$

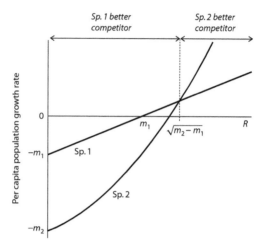

FIGURE 2.6. Permanent nonequilibrium coexistence in Levins's model (2.26). Inequality (2.30) requires that the curves describing the dependence of the per capita population growth rates of the two species on resource abundance intersect at a level of resource abundance $\sqrt{m_2 - m_1}$ that is higher than the value at which species 1 has zero per capita population growth rate, m_1. Thus, species 1 must be a superior competitor (i.e., its per capita population growth rate must be higher than that of species 2) at low resource abundance, and its per capita population growth rate must further be positive for some values of resource abundance before species 2 becomes the superior competitor. Species 2 is always the superior competitor at high resource abundance because of the nonlinear dependence of its per capita population growth rate on resource abundance.

Each nonlinearity in the per capita population growth rates introduces a statistical moment of the same order in the equations that define the constraints on coexistence and hence a new possibility of coexistence. It is possible to show that species 1 controls the temporal mean of the resource while species 2 reduces its temporal variance, so that species 1 may be viewed as a consumer of the mean while species 2 may be viewed as a consumer of the variance. Thus, the mean and the variance of resource abundance may be viewed as two distinct resources allowing the coexistence of two consumer species.

This counterintuitive result can be explained biologically as follows. The condition for permanent nonequilibrium coexistence [inequality (2.30)] requires that species 1 be a superior competitor at low resource abundance and that its per capita population growth rate be positive before species 2 takes over as the superior competitor (figure 2.6). Species 2 is always the superior competitor at high resource abundance because of the nonlinear dependence of its per capita population growth rate on resource abundance. Thus, while species 1 depresses and controls mean resource abundance, species 2 consumes more of the resource and grows faster during peaks of resource abundance, thereby reducing the variance of resource abundance. In other words, temporal variation in resource abundance allows temporal niche differentiation between the two species, which is the basis for their permanent nonequilibrium coexistence. This conclusion has since been confirmed both theoretically (Armstrong and MacGehee 1980; Huisman and Weissing 1999) and experimentally using algal microcosms (Sommer 1985).

Chesson (1994) formulated a more general discrete-time model of competition in variable environments and attempted to identify different mechanisms that may generate nonequilibrium coexistence. Defining the per capita population growth rate of species i during the time interval t to $t + 1$ as

$$r_i(t) = \ln N_i(t + 1) - \ln N_i(t), \tag{2.31}$$

he showed that the average per capita population growth rate of an invader at low density, $\bar{r}_i$, may be partitioned into a mean environmental effect, ΔE, a mean competitive effect, ΔC, and a mean of their interaction, ΔI:

$$r_i = \Delta E - \Delta C + \Delta I. \tag{2.32}$$

He identified two main mechanisms of permanent nonequilibrium coexistence using this model: (1) relative nonlinearity, which affects the mean competitive effect, ΔC; and (2) the storage effect, which affects the interaction between competition and the environment, ΔI. *Relative nonlinearity* captures the fact that different species have different nonlinear responses to competition. This is essentially what Levins showed with the above model. The *storage effect* captures the fact that the growth benefits afforded by favorable periods are stored to resist unfavorable periods of population decline. It requires that fluctuations in the environment and in the strength of competition covary and that they have antagonistic effects on per capita population growth rates, such that a species can have periods of exceptionally strong growth when competition is relatively low and environmental conditions are relatively good.

Both relative nonlinearity and the storage effect are generated by specific traits that determine the way species use resources or respond to their environment. As such, they may in principle be regarded as part of a species' niche. When the environment is periodic or seasonal, the correspondence with classical niche theory in a constant environment can be established explicitly. Temporal overlap and the difference between consumers' resource-use intensities play roles that are in general qualitatively similar to those of niche overlap and difference between carrying capacities, respectively, in the classical theory. However, the relative time scale of resource dynamics plays a crucial role that does not have its like in the classical theory: coexistence of consumers is possible only provided resource dynamics is sufficiently fast compared with consumer dynamics (Loreau 1992).

Since permanent coexistence in temporally fluctuating environments may in principle be understood using the conceptual tools of niche theory, its functional consequences should also be compatible with niche theory (Chesson et al. 2001). Levins's above model may again be used to explore this in the simplest case where species environmental responses are constant. In his model, species 1 sets mean resource abundance at m_1 irrespective of the presence of species 2. Introduction of species 2 further reduces the variance of resource abundance by consuming peaks of resource abundance. Thus, the two species are complementary in their resource use. The main effect of this temporal complementarity, however, is not on the mean but on the variance; i.e., the system fluctuates less. An additional effect on the mean should occur if the two species have differential responses to environmental fluctuations such as staggered seasonal activity patterns, in which case the combination of several species should maintain a lower level of resource throughout the year (Loreau 1992).

Transient Nonequilibrium Coexistence

Transient nonequilibrium coexistence occurs in all competitive systems since competitive exclusion is never instantaneous. Two very different theories appeal more particularly to transient coexistence to explain patterns of species diversity: the intermediate disturbance hypothesis and the neutral theory.

The *intermediate disturbance hypothesis* (Connell 1978; Huston 1979) is a direct application of Hutchinson's (1961) intuitive solution to his paradox of the plankton applied to finite time windows. If communities are subjected to periodic disturbances that are either very strong or very frequent, only those species that grow fast enough will be able to recover from these disturbances, and species diversity will be low. At the other extreme, if disturbances are either very weak or rare, slow-growing superior competitors will eventually exclude inferior species, and species diversity will also be low. The highest diversity will be achieved at intermediate levels of either intensity or frequency of disturbance, at which a transient balance will be established between the competitive advantages of fast-growing inferior competitors and slow-growing superior competitors. In the absence of the mechanisms that ensure permanent coexistence, all species but one may ultimately go extinct in such a system, but over finite time intervals an intermediate level of disturbance should maximize diversity. This intermediate level of disturbance, however, should depend on the average growth rate of the organisms considered, which sets the characteristic time scale of competitive exclusion in the system.

Chesson and Huntly (1997) presented an elegant critique of this theory by showing that the long-term dynamics of species relative abundances is independent of the intensity of competition and hence of disturbance, and that the only factor that can slow down competitive exclusion is the similarity of species competitive abilities. Delayed competitive exclusion at intermediate levels of disturbance occurred in Huston's (1979) model because periodic disturbances acted to decrease the intrinsic rates of natural increase and carrying capacities of all species to near-zero values, which is an unrealistic feature of his model. Another mechanism that can delay competitive exclusion in this model is a trade-off between r and K selection, which corresponds to differences in successional niches among species. This trade-off tends to equalize the competitive abilities of the various species at intermediate levels of disturbance. When species diversity is maintained by differences in successional niches, it does not have pronounced effects on the functional properties of the community as a whole because the functioning of a successional landscape is simply the average functioning of the successional stages present across the landscape (Kinzig et al. 2001).

Hubbell's (1979, 1986) *neutral theory* of coexistence is also a theory of transient nonequilibrium coexistence, at least in its initial formulation. This theory is based on the fact that deterministic competitive exclusion becomes infinitely slow as one approaches the limiting case where species are identical in all relevant respects, i.e., are ecologically equivalent. Hubbell argues that the unpredictability of the competitive environment experienced by each individual in species-rich plant communities is a selective factor for ecological equivalence. Coexistence then becomes neutral; the dynamics of the community is a random drift, just as in the neutral theory of molecular evolution. An isolated community would thus drift to extinction, but the time to extinction may be very large if the total size of the community is large. In the recent extended version of his theory, Hubbell (2001) explicitly considers factors that generate diversity and counterbalance this slow drift to extinction, i.e., immigration in a local community and speciation in the regional metacommunity (chapter 7). These factors yield a dynamical balance between species loss by extinction and species gain by immigration or speciation, very much as in MacArthur and Wilson's (1967) equilibrium theory of island biogeography. The nature of neutral theory has somewhat changed in the process since it has shifted from a theory focused on transient nonequilibrium coexistence to a theory focused on permanent nonequilibrium coexistence maintained by the regional processes of immigration and speciation. In this theory, species identity changes through time, but community properties such as the number of species and their abundance distribution reach a dynamical equilibrium maintained by the balance between extinction and immigration or speciation.

Neutral theory provides an elegant theoretical framework that drastically simplifies the description of multispecies communities and is able to reproduce a number of empirical patterns observed in plant communities, such as species abundance distributions and changes in community similarity as a function of distance. It is, however, highly controversial (Chave 2004; Holyoak and Loreau 2006; McGill et al. 2006b) because it questions the relevance of the huge amount of studies that have shown deterministic niche differences and competitive exclusion between species. As a matter of fact, it is the only theory that, in its current form, is incompatible with niche theory. A synthesis of the neutral and niche perspectives is, however, conceivable and desirable, just as is a synthesis of selection and neutrality in population genetics (Holyoak and Loreau 2006). Chapter 5 will present a nonneutral extension of neutral theory to examine the link between population- and community-level stability in multispecies communities. Neutral theory implies that species diversity has no functional consequence since all species and all combinations of species are functionally equivalent.

CONCLUSION: NICHE, COEXISTENCE, AND THE COMPETITIVE EXCLUSION PRINCIPLE

Niche and competition theory exerted considerable influence over the development of community ecology in the last decades of the 20th century. Sufficient empirical and experimental evidence has now been accumulated to show that competition is indeed a significant force in ecological systems under both laboratory and natural conditions. A very different question, however, is to know to what extent competitive exclusion and niche differentiation explain the composition, diversity, dynamics, and evolution of entire natural communities. As mentioned earlier, unambiguous patterns of niche differentiation driven by interspecific competition generally concern limited subsets of numerically dominant or ecologically related species within communities. The recent popularity of theories of coexistence that call upon spatial structure and nonequilibrum dynamics, especially for plants, shows that the niche differences predicted by classical competition theory in homogeneous environments are not always spectacularly obvious and prevalent.

There are essentially two ways to reconcile recent developments in the role of spatial structure and nonequilibrum dynamics in the maintenance of species diversity with classical niche theory. One option is to favor rigor and precision and adopt a restrictive view of the niche and of the competitive exclusion principle by explicitly limiting their scope to communities at equilibrium in homogeneous environments. The competitive exclusion principle could be thus formulated in such a restrictive framework: "There cannot be stable equilibrium coexistence between several species that are limited identically by the same resources in a homogeneous environment." This formulation could be easily extended to limiting factors other than resources as proposed by Levin (1970). In this framework, spatial structure and nonequilibrium dynamics are recognized as distinct factors that are not accounted for by the principle. This restrictive version would not harm the usefulness of the competitive exclusion principle since the latter is essentially an analytical principle that plays a role similar to Newton's first law of motion in classical mechanics, i.e., that of an ideal benchmark against which the action of natural forces can be contrasted (Hardin 1960).

Another option, however, is to favor generality and adopt a broader, more flexible view of the niche and of the competitive exclusion principle. As we have seen, permanent coexistence in spatially heterogeneous and temporally fluctuating environments has qualitatively similar constraints and functional consequences as does stable coexistence in homogeneous environments, i.e., niche differentiation and overyielding. Including species responses to environmental variations in space and time seems a natural extension of the niche concept. This extension, however, may be much less intuitive than it appears at first sight, as shown by Levins's (1979) example of permanent nonequilibrium coexistence on a single resource that hinges upon different species consuming different statistical moments of resource abundance. Even theories that call upon spatial structure created by the organisms themselves in an otherwise homogeneous environment or some of the theories based on transient nonequilibrium coexistence can be framed within niche theory since they require trade-offs between species traits, and these deterministic trade-offs can be viewed as different ways of using resources in space and time and hence as niche differences. Their functional consequences, however, are somewhat different since transgressive overyielding seems difficult to obtain in these cases. The only theory that is fundamentally incompatible with niche theory is neutral theory in its pure form since this assumes a complete absence of any deterministic difference among species that could be described as part of their niche.

In this broader framework, the competitive exclusion principle describes a limiting case characterized by three essential ingredients: (1) the identity of species environmental responses (which define their niches); (2) the nonidentity of species competitive abilities (as described by their carrying capacity, their R^*, or other such measures); and (3) asymptotic dynamical behavior (whether equilibrium or nonequilibrium). Coexistence is possible if at least one of these ingredients is not present, i.e., if species have different niches (classical niche theory and theories that call upon various trade-offs), if species are competitively equivalent (neutral theory), or if transient dynamical behavior is considered (theories that consider transient coexistence).

Given the variety of mechanisms that make coexistence possible even in purely competitive communities, it should come as no surprise that theory does not predict any absolute limit to the number of species that can coexist in a particular location based on species interactions alone. MacArthur's (1972) theory has sometimes been interpreted as implying that communities are saturated with species (Cornell and Lawton 1992; Cornell 1993). Given a fixed pool of species with fixed traits such as niche breadth and overlap, MacArthur's theory does predict that a single community emerges as a globally stable equilibrium configuration—the community that best utilizes available energy. But the composition and diversity of this equilibrium community are shaped by the particular species pool and resource availability distribution chosen. Although the species that coexist at equilibrium must be sufficiently dissimilar in the way they use resources, there is no absolute limiting similarity (Abrams 1983). Accordingly, communities may saturate through time for a given initial species pool, but they do not generally saturate with species with respect to changes in the species pool, as determined by evolutionary or historical processes (Loreau 2000a). As Whittaker (1972, p. 217) emphasized, "There is no evident intrinsic limit on the increase in species number, with increased packing and elaboration of axes of the niche hyperspace." A similar conclusion holds for any other coexistence mechanism. The number of species that a particular biotope harbors depends on its physical size (which limits the total number of individuals and hence the number of species), the environmental conditions in that biotope and its surroundings, and the multiple adaptations of species to the environmental conditions they have experienced in the past (including interactions with other species). A comprehensive theory that predicts species diversity from these basic environmental and evolutionary constraints is still missing.

The theory reviewed in this chapter suggests that some functional consequences of species diversity in competitive communities should be fairly general, though none is predicted to be universal. In particular, diverse communities should often show more efficient resource exploitation than do single species and hence overyielding. By contrast, transgressive overyielding is expected to occur only under specific coexistence mechanisms and requires large enough degrees of niche differentiation. These important issues are explored in more detail in the next chapter.

CHAPTER 3

Biodiversity and Ecosystem Functioning

During the last decade interest has shifted from explaining species diversity to understanding the functional consequences of biodiversity. Biodiversity is a broader concept than species diversity because it includes all aspects of the diversity of life—including molecules, genes, behaviors, functions, species, interactions, and ecosystems. Accordingly, it can be approached from multiple perspectives. Although the classical approach in taxonomy, ecology, and conservation biology has been based on species and species numbers, other approaches focus on the diversity of functional traits (Diaz and Cabido 2001; Naeem and Wright 2003) or phylogenies (Faith 1992). So far, however, most studies have concerned species diversity. As we saw in the previous chapter, community ecology has much to say on species diversity. But its focus has traditionally been on the processes that generate and maintain species diversity and has been largely theoretical or fundamental.

When ecologists and environmental scientists started to become interested in the relationship between biodiversity and ecosystem functioning, it was with a more practical viewpoint. Biodiversity is increasingly threatened by human activities and its environmental impacts, in particular, changes in land use, biological invasions, overexploitation of biological resources, pollution, and, more recently, climate change. Conservation efforts had mainly concerned individual species—most often, large, charismatic vertebrates because of their aesthetic value or for ethical reasons. But what about the multitude of other species that surround us? And what about the broader consequences of biodiversity loss? Living organisms drive energy flows and biogeochemical cycles in ecosystems from local to global scales: could the loss of their diversity affect the functioning of ecosystems and the "services" they deliver indirectly to humans?

When these important questions were first asked in the early 1990s (Schulze and Mooney 1993), neither community ecology nor ecosystem ecology was able to provide clear answers. There was much theoretical, experimental, and observational evidence for the importance of vertical diversity, in particular, for the critical roles played by top predators and keystone species in ecosystems, but little evidence and no consensus on the functional significance of horizontal diversity. Ecosystem ecology had largely ignored biodiversity, while community ecology was filled with competing theories about it, whose functional consequences were poorly understood and sometimes contradictory. The previous chapter showed that it is possible to bring order into this apparent proliferation of theories and predictions, but this theoretical housework had not been done at that time.

Given this state of affairs, a number of ecologists decided to address these issues experimentally using new approaches that manipulate entire ecosystems in the laboratory or in the field. Some of these experiments have been of considerable size, particularly those that have manipulated plant diversity in temperate grasslands. Theory has developed in the wake of these experiments and in close connection with them. The relationship between biodiversity and ecosystem functioning has been a very active and prominent research field at the interface between community and ecosystem ecology since then. It also provides a nice example of a successful interaction between theory and experiments in ecology. General syntheses of this research have been produced recently (Kinzig et al. 2001; Loreau et al. 2001, 2002b; Hooper et al. 2005; Naeem et al. 2009).

In this chapter, I seek to extract some of the most significant conceptual and theoretical advances that have been made in this research field. I focus on the relationship between horizontal diversity and ecosystem functioning within a single trophic level because it has been the main focus of recent experiments and it provides some simple principles that may form the basis for investigation of more complex systems. Therefore, this chapter is a logical continuation of the previous chapter. I address the more challenging issue of biodiversity and ecosystem functioning in complex systems with multiple trophic levels in the next chapter.

SMALL-SCALE EFFECTS OF BIODIVERSITY ON ECOSYSTEM FUNCTIONING

Most experiments about the effects of biodiversity on ecosystem functioning have manipulated the diversity of species or functional groups within a single trophic level on relatively small spatial and time scales. By "small scale", I mean here that the local environment within each plot is roughly homogeneous, so that spatial and temporal variations in environmental conditions can be ignored. Even the largest experiments performed during the last decade (Tilman et al. 2001; Spehn et al. 2005) are of this kind.

Their large size results from the large number of experimental plots they used, but they did not explicitly consider the effects of either spatial heterogeneity or temporal fluctuations in environmental conditions. A few recent experiments or analyses have begun to address spatial (Roscher et al. 2005) and temporal (Tilman et al. 2006) variability explicitly.

Biodiversity experiments have usually been designed to test the effects of diversity on ecosystem processes independent of variations in species composition and in any other factor that may drive biodiversity changes. Note that this is a restrictive view of biodiversity since the species composition of a community is one dimension of its diversity. But effects of changes in species composition due to the presence of species with strong functional impacts were already known in many instances, and the new intriguing question was whether diversity per se, independent of species identity, did matter for ecosystem functioning. To remove the effects of species identity and environmental factors, replicated communities at each diversity level were assembled by drawing species at random from a common species pool and grown under identical conditions at a given site.

Two of the largest such experiments have been the Cedar Creek biodiversity experiment in Minnesota, USA (Tilman et al. 1997a, 2001) and the BIODEPTH experiment in Europe (Hector et al. 1999; Spehn et al. 2005). Both experiments manipulated the diversity of plant species and the diversity of plant functional groups to test their effects on primary production and nutrient retention in temperate grassland ecosystems with a high level of replication. The advantage of BIODEPTH is that it was replicated over eight sites under different biogeographical, climatic, and soil conditions across Europe, which allows testing of the generality of biodiversity effects. The advantage of the Cedar Creek experiment is that it has been run for more than a decade, which allows testing of the robustness of biodiversity effects through time. The two experiments provided very similar results overall. BIODEPTH showed a log-linear increase in plant aboveground biomass production with species richness across sites (figure 3.1A). The Cedar Creek experiment showed a positive response of total plant biomass production to species richness, which became stronger through time (figure 3.1B). The number of plant functional groups also had positive effects on plant biomass in both cases. Since then a large number of biodiversity experiments have been performed, some of which, such as the Jena experiment, have surpassed the BIODEPTH and Cedar Creek experiments in terms of either plot size or replication level (Roscher et al. 2005). Not all of these experiments yielded significant results, but many did and provided results that generally followed the pattern in figure 3.1. Note, however, that

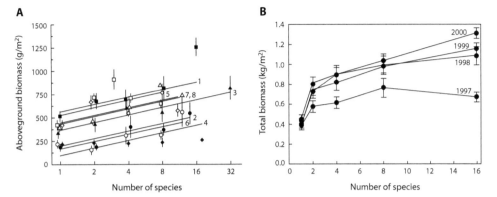

FIGURE 3.1. Effects of plant species richness on annual aboveground plant
biomass production in the BIODEPTH experiment (A) and on annual total
plant biomass production in the Cedar Creek experiment (B). Average biomass production increases with plant diversity across the eight sites in the
BIODEPTH experiment, an effect that becomes stronger through time in the
Cedar Creek experiment. Different symbols and lines correspond to different
sites in (A). Modified from Hector et al. (2002) and Tilman et al. (2001).

the response of plant aboveground biomass to species richness was non-
monotonic at two of the eight BIODEPTH sites when analyzed separately
(Spehn et al. 2005).

  These results generated a vigorous debate within the scientific community. Although they were consistent with simple expectations based on
niche theory, other theories, as well as previous results from replacement
series and intercropping experiments using low-diversity mixtures (Trenbath 1974; Vandermeer 1989), suggested that overyielding and niche differentiation might not be the rule in plant communities. Theoretical considerations indicated that two very different classes of mechanisms could
explain the relationship between average biomass production and initial
species diversity that was observed in plant biodiversity experiments.

  The first class of mechanisms involves either *niche differentiation* or *facilitation* between species, in agreement with niche theory (chapter 2). Models that incorporate niche differences along resource gradients or interspecific trade-offs between competitive abilities for different resources predict
an asymptotically increasing productivity as species diversity increases because of a better collective utilization of available resources in more diverse
competitive communities (Tilman et al. 1997b; Loreau 1998a). The relationship between productivity and diversity, however, depends on the nature and strength of species interactions in the community. A discrete-time

version of the classical Lotka–Volterra competition model can be used to examine this relationship under different scenarios of species loss (Gross and Cardinale 2005). Here I rewrite Gross and Cardinale's (2005) model in the form

$$N_i(t + 1) = N_i(t)\exp\left\{ r_m\left[1 - \frac{N_i(t) + \alpha\sum_{j \neq i} N_j(t)}{K'}\right]\right\}$$

$$= N_i(t)\exp\left\{ r_m\left[1 - \frac{(1 - \alpha)N_i(t) + \alpha N_T(t)}{K'}\right]\right\} \qquad (3.1)$$

In this equation, $N_i(t)$ is the population size of species i at time t, $N_T(t) = \sum_{j=1}^{S} N_j(t)$ is total community size at time t, S is the number of species, r_m is the instantaneous intrinsic rate of natural increase in all species, K' is their standardized carrying capacity, and α is their interspecific competition co-efficient. Following Gross and Cardinale (2005), I assume for simplicity that all species have equal intrinsic rates of natural increase, interspecific competition coefficients and carrying capacities. I also remove variations in initial community size before species loss by standardizing carrying capacities such that the equilibrium value of initial community size, K, is independent of α:

$$K' = \frac{1 + \alpha(S_0 - 1)}{S_0} K, \qquad (3.2)$$

where S_0 is the initial number of species.

The initial equilibrium abundance of each species, $N_{i,0}^*$, is K/S_0. Assume now that s species go extinct in the community described by equation (3.1). These extinctions reduce the number of species, S, from S_0 to $S_0 - s$ but do not affect their standardized carrying capacity as defined by equation (3.2). Therefore, their postextinction equilibrium abundance, $N_{i,s}^*$, is, from equation (3.1),

$$N_{i,s}^* = \frac{K'}{1 + \alpha(S_0 - 1 - s)} = \frac{K}{S_0}\left[\frac{1 + \alpha(S_0 - 1)}{1 + \alpha(S_0 - 1 - s)}\right] = \frac{K}{S_0}\left[1 + \frac{\alpha s}{1 + \alpha(S_0 - 1 - s)}\right].$$
$$\qquad (3.3)$$

Thus, their equilibrium abundance is increased by a proportional amount given by the second term in square brackets in equation (3.3). When the competition coefficient, α, is zero, their abundance remains constant because species do not interact and hence are unaffected by other species' extinction. The larger the competition coefficient, the larger the compensatory increase in the abundance of remaining species because of competitive release. A nice feature of this simple model is that it can also accommodate mutualistic interactions by letting α be negative. In this case, the equilibrium abundance of remaining species decreases by a proportional amount given by the second term in square brackets in equation (3.3) because they suffer from the loss of beneficial partners.

Gross and Cardinale (2005) refined the above model by allowing each species to have a different proneness to extinction and a different per capita productivity, yielding the following expression for the expected community productivity after s extinctions:

$$\Phi_s^* = \frac{K}{S_0}\left[1 + \frac{\alpha s}{1 + \alpha(S_0 - s - 1)}\right]\sum_i c_i(1 - p_{i,s}),\qquad(3.4)$$

where c_i is species i's per capita productivity, and $p_{i,s}$ is its cumulative extinction probability.

Equation (3.4) predicts that species loss leads to a linear decline in community productivity when $\alpha = 0$ because there is no compensation by remaining species, an accelerating decline when $0 < \alpha < 1$ because there is partial compensation by remaining species and the last species that go extinct have a large effect, and a decelerating decline when $\alpha < 0$ because the first species that go extinct have a beneficial effect on a large number of other species (figure 3.2). Looked at in the opposite direction, species richness is expected to yield a linear increase in productivity when $\alpha = 0$, a decelerating increase when $0 < \alpha < 1$, and an accelerating increase when $\alpha < 0$. In the limit when $\alpha = 1$, species richness has no effect on community productivity on average under random extinction because remaining species compensate exactly for the lost productivity of extinct species—until the last species is lost, of course, at which point community productivity vanishes. In the latter case, however, the order in which species are lost has a relatively large impact on the productivity–diversity relationship because different extinction scenarios yield qualitatively different shapes (figure 3.2). By contrast, extinction scenarios usually do not alter the qualitative shape of the productivity–diversity relationship when $\alpha < 1$. These predictions can be summarized as follows: niche differentiation and facilitation generate positive effects of species diversity on community productivity; the stronger niche differentiation and facilitation (i.e., the lower α), the stronger the enhancement of community productivity at high species diversity.

A second mechanism that could explain the relationship between average biomass production and initial species diversity in biodiversity experiments involves an effect of interspecific competition known as the *sampling effect* (Huston 1997; Tilman et al. 1997b). Assume that all species compete for a single limiting resource without any niche difference between them.

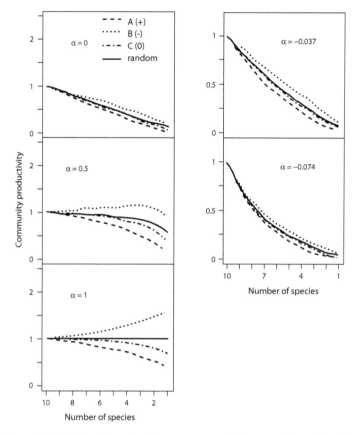

FIGURE 3.2. Effects of species richness on expected community productivity in a discrete-time Lotka-Volterra competition model for various values of the competition coefficient, α, and various scenarios of species extinction. Solid lines: random extinction (extinction risk equal among species). Dashed lines: extinction risk ordered by per capita productivity (scenario A). Dotted lines: extinction risk in reverse order of per capita productivity (scenario B). Dotted-dashed lines: extinction risk differs among species but is unrelated to per capita productivity (scenario C). Negative values of the competition coefficient, α, correspond to mutualistic interactions between species. Modified from Gross and Cardinale (2005).

Eventually, a single species outcompetes all the others—the species that has the lowest equilibrium resource requirement (chapter 2). This species also has the highest biomass and the highest productivity under some conditions that were identified in chapter 2. Now suppose that species are drawn randomly from a pool to assemble communities in each plot, as was done in a number of biodiversity experiments. The probability of sampling

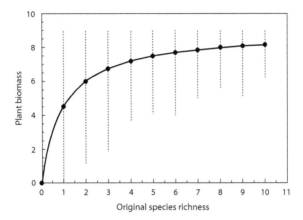

FIGURE 3.3. The sampling effect: plant biomass is expected to increase, on average, asymptotically with initial species richness, the asymptote being determined by the biomass of the most competitive species in the species pool. Modified from Tilman et al. (1997b).

a highly productive species from the pool increases with the number of species sampled to assemble a community. As a result, total productivity and total biomass are expected to increase, on average, asymptotically with initial species richness, the asymptote being determined by the productivity or biomass of the most competitive species in the pool (figure 3.3). Note that, in this scenario, all communities eventually become monocultures. Thus, species diversity is not maintained through time; it determines only the initial range of trait variation from which the competitively dominant species is drawn. That is why the sampling effect generated so much debate. Some (Huston 1997; Aarssen 1997; Wardle 1999) interpreted it as a statistical artefact or hidden treatment invalidating biodiversity experiments, whereas others (Tilman 1997; Tilman et al. 1997b) interpreted it as a valid biodiversity effect. I shall come back to this issue later.

Theory was instrumental in resolving this controversy over potential mechanisms that have vastly different implications. Scientific controversies are often the result of a lack of clarity in the theoretical framework, a lack of appropriate tools, or a lack of sufficient empirical evidence to distinguish among clearly identified competing hypotheses. This one was no exception and combined all three problems. Therefore, the first step in its resolution was a theoretical advance in the conceptual framework in which the experiments were being conceived and interpreted. The debate was plagued by at least three conceptual problems: (1) the idea that species richness per se was responsible for the observed effects of diversity since it was the experimentally manipulated variable; (2) the belief, based on existing simple theory, that species diversity should necessarily have a positive effect, if any, on productivity; and (3) a dichotomous view of probabilistic sampling processes and deterministic niche processes.

My theoretical work using a mechanistic ecosystem model (Loreau 1996, 1998a) challenged these views. My model described plant competition for a limiting soil nutrient in a spatially structured environment as discussed in chapter 2. The particular mechanism involved, however, is not critical. The main features of the model that made progress possible are that (1) it is fundamentally an individual-based, if highly simplified, model that does not prescribe any a priori grouping by species or functional group; (2) it is both simple enough to be tractable analytically and realistic enough to be applicable to actual experiments; and (3) it is flexible enough to explore a variety of scenarios that combine sampling and niche processes. Species richness was chosen as a measure of biodiversity to mimic experiments, but it is really phenotypic diversity as determined by intra- or interspecific differences in plant traits that matters in this model. Two plant traits turned out to play a critical role: (1) the potential overlap between their individual resource depletion zones, which determines niche differences in space occupation among species, and (2) their resource-use intensity, i.e., their ability to depress the resource level locally, which determines their competitive ability.

Using this model, I explored two limiting cases of potential niche overlap in space occupation. One limiting case occurs when plants from all species have the same root architecture and potentially occupy identical resource depletion zones. I called such species redundant because they occupy the same spatial niche and thus fulfill the same functional role, even though their competitive ability may differ. The other limiting case occurs when plants from different species have very different root architectures and occupy completely nonoverlapping spaces. I called such species complementary because they occupy distinct spatial niches and thereby fulfill complementary functional roles. When species are redundant, the effect of biodiversity on primary productivity, total plant biomass, and nutrient retention depends on the way average competitive ability varies with diversity. Ecosystem properties do not respond to changes in species diversity if mean competitive ability does not covary with diversity, but they can either increase or decrease if there is some selective process that makes mean competitive ability increase or decrease with diversity. This case is equivalent to the case where $\alpha = 1$ in model (3.1), and the predicted patterns look very much like those obtained for the various extinction scenarios explored in the bottom left panel in figure 3.2 because the latter are effectively different ways to let mean competitive ability vary with diversity. By contrast, when species are complementary, primary productivity is expected to always increase asymptotically with diversity, and changes in mean competitive ability affect only the steepness of the response. This case is equivalent to the case where $0 < \alpha < 1$ in model (3.1), and the predicted patterns are similar to those of the middle left panel in figure 3.2.

This theoretical work showed that there are two types of mechanisms by which biodiversity influences productivity or other ecosystem processes in competitive systems, leading to two types of biodiversity effects: (1) functional niche complementarity—the *complementarity effect*, and (2) selection of particular functional traits that affect species' competitive abilities—the *selection effect* (Loreau 2000b). In both cases, biodiversity provides a range of trait variation that is the raw material for the operation of these effects. In the complementarity effect, trait variation forms the basis for a permanent association of species that enhances collective performance. In the selection effect, trait variation comes into play as an initial condition, and a selective process then promotes dominance by species with particular functional traits. This apparently straightforward conceptual clarification was instrumental in resolving the debate over biodiversity experiments in several ways.

First, it became clear that species diversity, or biodiversity at large, matters only for ecosystem functioning to the extent that it provides *phenotypic trait variation* related to the particular ecosystem process considered. Recent experiments have manipulated species richness as a surrogate for functional trait diversity, which is much more difficult to measure a priori, especially when different ecosystem processes are studied simultaneously. But there is no magic effect of the number of species per se. What matters is functional trait diversity within the community, whether this diversity is among individuals of the same species (genetic or phenotypic intraspecific diversity), among individuals of different species (species diversity), or among individuals of different functional groups (functional-group diversity). New approaches based on functional diversity have now been developed (Heemsbergen et al. 2004; Petchey and Gaston 2002; Wright et al. 2006).

Second, it also became clear that the sampling effect is only a special case of a more general mechanism. In fact, the sampling effect as originally proposed has two independent components: (1) a probabilistic sampling component—sampling more species means a higher probability of sampling species with particular traits, such as a high intrinsic productivity; and (2) a deterministic selection component—the most productive species is favored by competition and becomes dominant to the point of excluding other species. Paradoxically, its sampling component is not what distinguishes the sampling effect. Sampling effects occur irrespective of the strength of competition and selection. Sampling more species allows sampling a wider range of functional traits, and this wider range of functional traits is the basis for both the complementarity and selection effects. As for the selection component of the sampling effect, it is but a special, extreme case of the selection effect. Complete dominance of the most productive species leading to the complete or virtual elimination of inferior competitors is not required for a positive selection effect to be present in which higher-yielding species are favored. In fact, negative selection effects are possible if competitive ability is correlated negatively with productivity because of trade-offs between resource acquisition and interference competition, or between resource acquisition and resource-use efficiency. The more general and flexible concept of the selection effect allows it to be reconciled with the complementarity effect. Selection and complementarity are likely to operate in combination in any concrete situation and may be viewed as two endpoints along a continuum from pure selection to pure complementarity (figure 3.4). Intermediate situations are characterized by complementarity between particular species or functional groups, or, equivalently, by selection of particular combinations of species or functional groups. The realization that selection and complementarity are complementary, not contradictory, processes contributed to build a consensus conceptual framework on biodiversity and ecosystem functioning (Loreau et al. 2001).

Third, there is an obvious analogy between the ecological selection effect and the evolutionary process of natural selection. This analogy suggests that the selection effect should not be discarded as purely artefactual. It is true that the two types of mechanisms identified above have different implications for the significance of biodiversity for ecosystem functioning. Functional complementarity arises from and maintains biodiversity, is predictable from the individual species' biological traits, and generally has a consistent, positive effect on several ecosystem processes, such as total plant biomass, primary production, and nutrient retention. The selection effect contributes to erode biodiversity (at least in a constant environment), it is likely to be more variable (in particular, it can be positive or negative), and its significance may vary depending on a number of factors, including the kind of systems, the kind of organisms, and the spatial and temporal scale considered, as well as the scientist's or the manager's objectives. Managed or agricultural ecosystems, for instance, experience artificial selection by humans, and accordingly natural selection effects may have little relevance for those who manage them in the short term. But the assembly (species gain) or disassembly (species loss) of natural ecosystems has a

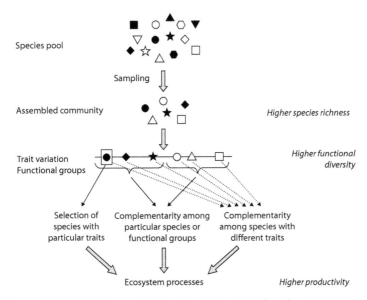

FIGURE 3.4. Mechanisms underlying the effects of biodiversity on ecosystem processes in controlled biodiversity experiments. Sampling effects are involved in community assembly, such that communities that have more species have a greater probability of containing a higher phenotypic trait diversity. Phenotypic diversity then maps onto ecosystem processes through two main mechanisms: selection of species with particular traits, and complementarity among species with different traits. Intermediate scenarios involve complementarity among particular species or functional groups or, equivalently, dominance of particular subsets of complementary species. Modified from Loreau et al. (2001).

component of chance or historical contingency. The predicted global changes in climate, land use, and biological invasions will likely have huge effects on the local dominance and proneness to extinction of species and greatly increase this component of contingency. Also, many systems in the world are harvested rather than managed. Harvesting often selectively removes species with dominant or key roles and may drive them to extinction—fisheries provide a good example of this threat. Thus, the performance of species and the order in which they are likely to be lost in the future might be highly variable and unpredictable. Finally, selection at one scale may turn into complementarity at another scale. Selection of the best-adapted species at any local site generates complementarity between species across sites if the environment is spatially heterogeneous (Chesson et al. 2001). Similarly, selection of the best-adapted species at any point in time may play a critical role in the long term in temporally fluctuating environments (chapter 5). All this argues for considering the selection effect as a relevant biological effect, albeit with different implications than the complementarity effect.

IDENTIFYING MECHANISMS IN
BIODIVERSITY EXPERIMENTS

After clarification of the conceptual and theoretical framework in which biodiversity experiments can be properly interpreted, the second step in the resolution of the controversy over these experiments consisted in devising new theoretical tools to analyze their results. Since several experiments had already been completed at a considerable cost, how could the data collected in these experiments be used to assess the respective contributions of the selection and complementarity effects? To do this, the conceptual distinction between the selection and complementarity effects had to be made operational. Andy Hector and I developed a new methodological approach to separate the two effects based on an *additive partition of biodiversity effects* analogous to the Price equation in evolutionary genetics (Loreau and Hector 2001). Our methodology builds both on the analogy between the selection effect and natural selection in evolutionary biology, and on the concept of relative yield developed in agricultural sciences (chapter 2).

Concretely, first define the net biodiversity effect of a mixture as the difference between its observed yield and its expected yield under the null hypothesis that there is no selection or complementarity. This expected value is the weighted (by the relative abundance of species in the mixture) average of the monoculture yields for the component species. The selection effect is then measured by the covariance between the monoculture yield of species and their change in relative yield in the mixture. On this definition, positive selection occurs if species with higher-than-average monoculture yields dominate mixtures, while negative selection occurs if species with lower-than-average monoculture yields dominate mixtures. Last, a positive complementarity effect occurs if species yields in a mixture are on average higher than expected based on the weighted average monoculture yield of the component species. Thus, complementarity is defined operationally as any increase in the performance of communities above that expected from the performance of individual species and includes the effects of both niche differentiation and facilitation between species. Indeed, distinguishing the effects of niche differentiation and facilitation may often be difficult in practice. For instance, one common form of complementarity in plant communities comes between legumes, which have the ability to fix atmospheric nitrogen, and other plants, which have access only to soil nitrogen. In this interaction, it is difficult to separate resource partitioning (the two functional groups use partly different nitrogen resources) from facilitation (legumes may have an indirect facilitative effect on other plants by bringing additional nitrogen into the ecosystem). Note that, just like the selection effect, the complementarity effect can also be negative. This occurs when interference competition between species prevails over niche differentiation and facilitation and reduces the performance of communities below that expected from the performance of individual species (Loreau 1998c).

The various biodiversity effects are then related by additive partition as follows (Loreau and Hector 2001):

$$\Delta Y = S.\overline{\Delta RY}.\overline{M} + S.\mathrm{cov}(\Delta RY, M). \tag{3.5}$$

In this equation, ΔY measures the net biodiversity effect, $S.\overline{\Delta RY}.\overline{M}$ measures the complementarity effect, $S.\mathrm{cov}(\Delta RY, M)$ measures the selection effect, and the following definitions hold:

M_i = yield of species i in monoculture;
$Y_{O,i}$ = observed yield of species i in the mixture;
$Y_0 = \Sigma_i Y_{0,i}$ = total observed yield of the mixture;
$RY_{E,i}$ = expected relative yield of species i in the mixture, which is simply its proportion seeded or planted;
$RY_{O,i} = Y_{O,i}/M_i$ = observed relative yield of species i in the mixture;
$Y_{E,i} = RY_{E,i}M_i$ = expected yield of species i in the mixture;
$Y_E = \Sigma_i Y_{E,i}$ = total expected yield of the mixture;
$\Delta Y = Y_O - Y_E$ = deviation from total expected yield in the mixture;
$\Delta RY_i = RY_{O,i} - RY_{E,i}$ = deviation from expected relative yield of species i in the mixture;
S = number of species in the mixture.

When we applied this methodology to patterns of aboveground biomass production obtained in the BIODEPTH project, we found that the net biodiversity effect was positive overall (the grand mean was significantly different from zero) and increased significantly with species richness beyond two species; the selection effect was variable, ranging from significantly positive averages in two localities to a significantly negative average in one locality, but was not significantly different from zero across sites; and the complementarity effect was significantly positive overall and increased significantly with species richness beyond two species (figure 3.5). Thus, the

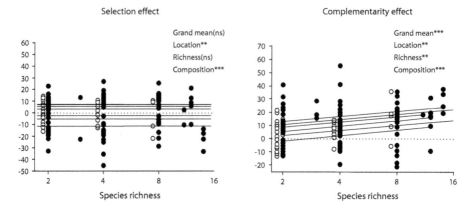

FIGURE 3.5. Selection and complementarity effects on aboveground plant biomass production as functions of species richness in the BIODEPTH experiment. Open circles are plots that do not contain any legume species; filled circles are plots that contain legumes. Lines are slopes from the multiple regression
model using species richness on a $\log_2$ scale for the eight sites. Values of biodiversity effects (in g/m²) are square-root-transformed to meet the assumptions
of analyses but preserve the original positive and negative signs. *: $P < 0.05$,
**: $P < 0.01$, ***: $P < 0.001$. Modified from Loreau and Hector (2001).

complementarity effect was stronger and more predictable than the selection effect, as expected from theory, and explained the overall pattern of
increasing biomass production with increasing plant species richness.

Since then, a large number of studies have used our additive partitioning equation to analyze the results of other biodiversity experiments. When
a positive biodiversity effect was present, it was often predominantly driven
by complementarity, not selection (Cardinale et al. 2007). As a matter of
fact, negative selection effects are relatively common, which shows that the
most productive species do not usually dominate multispecies communities, thus invalidating the hypothesis behind the sampling effect. Note that
our partition is not formally identical to the Price equation. In particular,
relative yields do not add up to 1 when there is complementarity (contrary
to relative frequencies in the Price equation), which tends to inflate the selection effect. Fox (2005) proposed a tripartite additive partition that di-
vides the selection effect into two parts to remove this problem. It remains
to be seen whether his new method will bring significant new insights in the
analysis and interpretation of biodiversity experiments.

Our additive partitioning methodology contributed to largely resolve
the controversy over the sampling effect in biodiversity experiments. Although the debate resurfaces at times, applying this methodology is usually

the fastest and most efficient way to settle it. For instance, Cardinale et al. (2006) recently performed a meta-analysis of a large number of biodiversity experiments and found that consumer biomass and resource use generally increased asymptotically with species richness in all types of organisms and at all trophic levels, but that transgressive overyielding was uncommon. They interpreted this result as evidence for the sampling effect based on the erroneous expectation that complementarity and coexistence between species should result in transgressive overyielding. As chapter 2 showed, complementarity and coexistence between species should result in overyielding but not necessarily in transgressive overyielding. Therefore, absence of transgressive overyielding cannot be taken as evidence for absence of complementarity. Applying the additive partitioning equation to those studies where the necessary data were available allowed us to jointly conclude that complementarity is in fact common in biodiversity experiments (Cardinale et al. 2007). We also found that complementarity increases significantly with the duration of the experiment, which provides evidence that it is not a transient effect generated by the perturbations inevitably associated with the establishment of experiments.

Despite its considerable value, the additive partitioning methodology has limitations, of course. In particular, detecting a significant complementarity effect does not tell how many species play a significant part in it, hence where we stand along the continuum from pure selection to pure complementarity in figure 3.4. Unfortunately, there is no theory or methodology that allows answering this question with certainty based on the data collected in recent biodiversity experiments. Our analysis of the BIODEPTH data rejected the hypothesis that the well-known complementarity between nitrogen-fixing legumes and other plants is sufficient to explain the positive complementarity effect found in this experiment since species richness retained a significant log-linear effect on complementarity when the presence of legumes was included as an additional factor in our across-site analyses (figure 3.5). Tilman et al. (2001) provided circumstantial evidence that 9 to 13 species contributed to the increase in plant biomass with diversity in their experiment, although 4 species played a dominant role. Hector and Bagchi (2007) used the Akaike information criterion to identify the most parsimonious set of species that are likely to influence an ecosystem process and found that on average 3 to 7 species affected a single ecosystem process in the BIODEPTH experiment, but that this number increased to 8 to 18 species for multiple ecosystem processes. The strongest evidence to date for a complementarity effect that goes beyond the classical complementarity between grasses and legumes in temperate grasslands comes from an experiment performed by

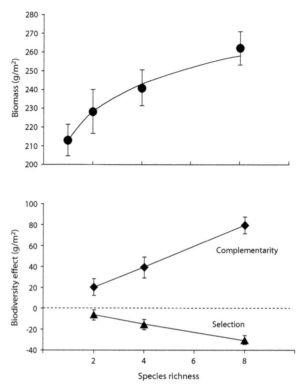

FIGURE 3.6. Effects of plant species richness on annual aboveground plant biomass production (top) and the corresponding selection and complementarity effects (bottom) in an experiment without legumes. Modified from van Ruijven and Berendse (2003).

van Ruijven and Berendse (2003), who excluded legumes altogether. Yet they found an increase in plant biomass, a positive increasing complementarity effect, and a negative decreasing selection effect with plant species richness, just as in previous experiments that included legumes (figure 3.6).

Another limitation of the additive partitioning methodology is that the selection and complementarity effects describe only broad types of mechanisms; they do not provide insights into the specific biological mechanisms that generate them. There is no miracle: specific biological mechanisms can be detected only by directly studying their specific signatures. Few experimental studies have explored the detailed biological mechanisms that underlie selection and complementarity effects. Differences in rooting depths and architectures (Dimitrakopoulos and Schmid 2004), increased nutrient use efficiency (van Ruijven and Berendse 2005), and increased input and

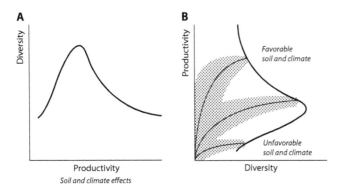

FIGURE 3.7. Relationships between (A) diversity–productivity patterns driven by environmental conditions across sites, and (B) the local effect of species diversity on productivity. (A) Comparative data often indicate a unimodal relationship between diversity and productivity driven by changes in environmental conditions. (B) Experimental variation in species richness under a specific set of environmental conditions produces a pattern of decreasing between-replicate variance and increasing mean response with increasing diversity, as indicated by the thin, curved regression lines through the scatter of response values (shaded areas). Modified from Loreau et al. (2001).

retention of nitrogen (Fargione et al. 2007) are some of the mechanisms that have been shown to be involved in functional complementarity between plant species in temperate grasslands. But other forms of resource partitioning and facilitation are likely to operate in these and other communities. Cardinale et al. (2002) provided a nice example of complementarity between caddisfly larvae in stream mesocosms where enhanced resource consumption by the insect community is due to hydrodynamic facilitation between different species through alterations in near-bed water flow.

RECONCILING THE RESULTS OF SMALL-SCALE EXPERIMENTS AND LARGE-SCALE COMPARATIVE STUDIES

The relationship between productivity and diversity has long been studied from a different angle than that in recent experimental studies, i.e., in the form of correlations across different sites or nutrient addition treatments. Although such diversity–productivity relationships show considerable variation across spatial scales and ecosystems (Waide et al. 1999; Gross et al. 2000; Mittelbach et al. 2001), a relatively common pattern in plants at geographical scales smaller than continents is a hump-shaped curve when species richness is plotted against surrogates of productivity (figure 3.7A)

(Mittelbach et al. 2001). Some comparative approaches have also suggested negative relationships between plant species evenness and rates of various ecosystem processes (Wardle et al. 1997). The differences between these large-scale, observational approaches and the small-scale experimental approaches have also generated debate (Tilman et al. 1997c). How can these apparently contradictory results be reconciled?

The two approaches examine different causal relationships under different sets of conditions. The traditional observational approach seeks to identify the *causes of spatial variation in diversity across environmental gradients*. Variation in diversity is often correlated with productivity and also with many other factors that influence productivity, such as soil fertility, climate, disturbance regime, and herbivory. By contrast, the recent experimental approach examines the *specific effects of diversity on productivity* within each site when all these other factors are held constant. The two approaches can be reconciled by considering that comparative studies reveal correlations between diversity and productivity driven by environmental factors, whereas small-scale experiments reveal the effects of species properties and diversity on productivity that are detected after the effects of other environmental factors have been removed (Loreau et al. 2001).

Here again, theory played a useful role by identifying the potential interactions between environmental factors, biodiversity, and productivity. I used the same mechanistic nutrient-limited ecosystem model as above to analyze how variations in environmental factors simultaneously affect plant diversity and primary productivity to mimic across-site comparisons (Loreau 1998a). Assume that, at each site, species diversity reaches an equilibrium level determined by local environmental constraints and the distribution of traits available in the regional species pool. Further assume that species are complementary, in which case the effect of diversity on productivity is positive at each site. Variations in abiotic factors among sites then have two effects on productivity: a direct effect, and an indirect effect through changes in species diversity (figure 3.8, right panels). The analysis of my model showed that the direct effect almost always prevails over the indirect effect. When the two effects are convergent, as is the case when soil fertility varies, a positive correlation between productivity and diversity emerges across sites (figure 3.8A). But when the two effects are conflicting, as is the case when the rate of soil nutrient diffusion varies, a negative correlation emerges across sites (figure 3.8B). When several abiotic parameters vary simultaneously (here, both soil fertility and the rate of soil nutrient diffusion), the correlation between productivity and diversity across sites tends to disappear (figure 3.8C) or can be combined to produce a hump.

Yet in all cases controlled experiments would reveal that diversity has an intrinsic positive effect on productivity at each site. The same results hold for total plant biomass and nutrient retention. Thus, the effects of environmental parameters on ecosystem processes mask the local effect of diversity on these processes in across-site comparisons.

This is an important but disturbing conclusion. One implication of this conclusion is that comparisons across sites cannot reveal the short-term effect of diversity on ecosystem processes unless abiotic conditions are very tightly controlled, which severely restricts the use of such comparisons. Another implication of this conclusion is that, when species diversity has the time to adjust to abiotic changes in the environment, the impact of changes in diversity on ecosystem functioning is expected to be secondary compared with the direct impact of environmental changes. In the case of gradual environmental changes such as changes in climate, for instance, their impact on ecosystem functioning might be approximately predicted by ignoring the dynamics of species diversity. This should not be true, however, for the massive species losses predicted in the future. In the latter case, changes in biodiversity may play a significant independent role, which cannot be ignored.

On the other hand, the above theory shows nicely that documented cases of negative or absent relationships between productivity and diversity across sites are entirely compatible with positive local effects of diversity on productivity. As a matter of fact, the data on aboveground plant biomass production from the BIODEPTH experiment matches theoretical predictions fairly well. Despite large differences in primary productivity between locations and no apparent relationship between productivity and maximum within-site species richness, productivity generally declined as species were lost within a site (figure 3.1A), generating a pattern that looks very much like figure 3.8C when several abiotic parameters vary simultaneously.

One may therefore reconcile the results from small-scale experiments and large-scale comparative studies by inverting the classical diversity–productivity plots of comparative studies (figure 3.7A) to produce a productivity–diversity relationship that corresponds to the envelope constrained by environmental factors into which diversity can vary. Within this envelope, diversity is expected to decrease productivity if all other factors are held constant, as predicted by theory and confirmed by experiments (figure 3.7B).

One way to avoid some of the problems involved in across-site observational studies is to control statistically for as many environmental factors as possible when analyzing comparative data. *A posteriori statistical control*

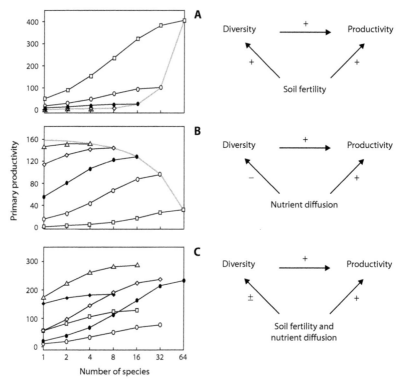

FIGURE 3.8. Within-site versus across-site relationships between primary productivity and plant diversity predicted by a mechanistic nutrient-limited ecosystem model (left), and causal relationships between abiotic environmental factors, plant diversity, and primary productivity that generate these predicted relationships (right). Species are assumed to be complementary; therefore species diversity always has a positive effect on primary productivity. Thin lines in the left panels show the positive effect of species richness on primary productivity within each site (i.e., holding environmental parameters constant) as controlled experiments would reveal it. The various lines correspond to different sites defined by different sets of environmental parameters. The right endpoint on each line corresponds to the maximum number of species that can coexist at equilibrium at each site based on a regular distribution of species competitive abilities; this is assumed to be the "natural" situation at each site. Across-site comparisons would reveal a positive (A, thick gray line), a negative (B, thick gray line), or no (C) relationship between primary productivity and plant diversity, depending on the environmental parameters that vary among sites and regardless of the positive effect of species diversity on primary productivity within each site. In (A), sites differ in an environmental parameter—here, the inflowing nutrient concentration in the soil, which determines soil fertility—that has positive direct effects on both diversity and productivity. This generates the positive correlation between productivity and

yields much weaker inference than a priori control through experimental design because it is impossible to be sure that one has taken all the important environmental variables into account, and these variables can have highly nonlinear effects. But weak inference is better than no inference at all. Comparative studies have great value in revealing patterns that would otherwise remain out of the reach of experimental studies when these patterns involve large spatial scales, long time scales, or not very accessible ecosystems. Controlling statistically for covarying environmental factors is then a useful approach to help formulate reasonable hypotheses about underlying causation and mechanisms. This approach has been followed in several recent studies (Grace et al. 2007; Vila et al. 2007).

Danovaro et al.'s (2008) work is a good example of the benefits that can be gained from such an approach (Loreau 2008). Deep-sea ecosystems are the most extensive ecosystem type on Earth's surface; they host a large fraction of Earth's biodiversity (most of it yet undiscovered), and yet they remain largely terra incognita because of their limited accessibility. Danovaro et al. (2008) used comparative data on a large number of deep-sea ecosystems across the globe to show that the relationship between species diversity and a number of ecosystem properties is exponential on the ocean floor. They also provided circumstantial evidence that covarying environmental factors are unlikely to fully explain this relationship because the latter still held after controlling for temperature, water depth, and carbon inflow from the photic zone, three of the main environmental factors that vary across deep-sea ecosystems. Although these statistical analyses are not sufficient to completely rule out any influence of covarying factors, this study is the first to suggest a nonsaturating relationship between productivity and diversity compatible with mutualistic interactions (figure 3.2, right panels). Given the expanse of deep-sea ecosystems and their potential impact on global biogeochemical cycles, these are challenging results.

---

diversity across sites. In (B), sites differ in an environmental parameter—here, the rate of nutrient diffusion between resource depletion zones and the global soil nutrient pool—that has a positive direct effect on productivity but a negative effect on diversity. Since the positive direct effect of the environmental factor on productivity prevails over the negative indirect effect through changes in diversity, this generates a negative correlation between productivity and diversity across sites. In (C), the two parameters are varied simultaneously, yielding no consistent relationship between productivity and diversity across sites. Modified from Loreau (2000b).

CONCLUSION

Strong interactions between theory and experiments have made significant advances possible on the impacts of biodiversity on ecosystem functioning during the last decade. Thanks to this combination of theory and experiments, we can safely conclude today that biodiversity does have the potential to affect ecosystem functioning to a measurable extent, even in the simple systems and on the small spatial and temporal scales considered in recent experiments, and that some form of complementarity between species driven by niche differentiation or facilitation is responsible for these effects. How many species play a significant role in these effects, which ecosystem processes are affected, and under what conditions, however, are largely empirical questions that are beyond the scope of the present book. I shall continue to explore these issues from a theoretical perspective in the following chapters by considering more complex and more realistic ecosystems that have several trophic levels, change through time, or vary in space.

Whether biodiversity loss will affect large-scale patterns of productivity hinges on the shape and steepness of the local dependence of productivity on diversity. Generally speaking, the relative effects of individual species and species richness may be expected to be greatest at small to intermediate spatial scales, while these biological factors should be less important as predictors of ecosystem processes at regional scales where environmental heterogeneity is greater. While diversity was manipulated as the independent variable in recent experiments, at large scales species diversity itself is a dynamical variable and adjusts to changes in environmental conditions. Theory then suggests that abiotic factors should be the main drivers of variations in ecosystem processes across environmental gradients. The dynamics of biodiversity and ecosystem processes at large scales, however, involves complex spatial processes that can strongly affect these conclusions. I shall return to this issue in chapter 7.

Food Webs, Interaction Webs, and Ecosystem Functioning

A food web describes the network of trophic interactions between species, i.e., who eats whom, in an ecosystem. Since trophic interactions are both the vehicle of energy and material transfers and one of the most significant ways in which species interact, they have always lain at the confluence of community and ecosystem ecology. But they have been approached from different perspectives in different traditions. The energetic view articulated by Lindeman (1942) and developed by ecosystem ecology during the following decades views food webs as networks of pathways for the flow of energy in ecosystems, from its capture by autotrophs in the process of photosynthesis to its ultimate dissipation by heterotrophic respiration. A different approach, rooted in community ecology, was initiated by Elton (1927) and developed by May (1973), Pimm (1982), and many others. This approach focuses on the dynamical constraints that arise from species interactions and emphasizes the fact that too much interaction (whether in the form of a large number of species, a large connectance among these species, or a high mean interaction strength) destabilizes complex ecological systems, including food webs. Food webs have also been studied from a topological perspective: the pattern of trophic interactions in a food web is nonrandomly related to species traits, in particular, body size, which led to the development of size-based models of food-web structure such as the cascade and niche models (Cohen et al. 1990; Williams and Martinez 2000). Perhaps the approach that lies most closely to the interface between community and ecosystem ecology is that based on the trophic cascade concept (Carpenter et al. 1985). Hairston et al. (1960) hypothesized that carnivores control herbivores, thereby releasing plants from control by herbivores, in most ecosystems. This simple idea led to a flurry of studies on the community- or ecosystem-level consequences (though mostly the biomass of the various trophic levels) of the top-down control exerted by higher trophic levels on lower trophic levels.

These different approaches remain poorly integrated with each other. They have also remained largely separated from the recent development of the biodiversity–ecosystem functioning area. As we saw in the previous chapter, most of the recent theoretical and experimental studies on the effects of biodiversity on ecosystem functioning have considered single trophic levels, primary producers for the most part. Although they have contributed to merging community and ecosystem ecology, they have unintentionally disconnected the vertical and horizontal dimensions of biodiversity and ecosystem processes. An important current challenge is to understand how trophic interactions affect the relationship between biodiversity and ecosystem functioning (Duffy et al. 2007). Several experimental and theoretical studies have started to investigate this issue, but the challenges, particularly for theory, are still considerable. Integrating the horizontal and vertical dimensions of biodiversity and ecosystems requires merging the approaches of competition theory (which is concerned with the maintenance of biodiversity within a simple trophic level) and food-web theories (which are concerned with the topological, dynamical, and functional properties of interaction networks). It is high time to lay a bridge between these approaches to foster cross-fertilization and build a broader theoretical framework that has greater relevance to natural ecosystems.

Even more challenging is the need to incorporate nontrophic interactions in ecological theory. Examples of mutualistic interactions (Bronstein 1994), ecosystem engineering (Jones et al. 1994), and trait-mediated indirect interactions (Schmitz et al. 2004; Werner and Peacor 2003) abound in nature. Yet they are most often ignored by ecological theory. One reason for this state of affairs is that simple models do not realistically describe these interactions. The Lotka–Volterra model of mutualism (Gause and Wit 1935), for instance, leads to unlimited population growth "in an orgy of mutual benefaction" (May 1981) when the interaction is too strong. Population explosions due to mutualistic interactions are frequent in random model interaction webs. These unrealistic features occur because the simple models used in community ecology ignore mass-balance constraints arising from the physical law of mass conservation.

In this chapter I extend the theory presented in the previous two chapters to more complex ecosystems that have multiple trophic levels connected by both trophic and nontrophic interactions. I first revisit some of the basic properties of trophic interactions and their consequences for ecosystem functioning by bringing together the energetic approach that focuses on energy flows and efficiencies and the dynamical approach that underlies the trophic cascade. I derive some new predictions about patterns of productivity and ecological efficiency in food chains and show that trophic interactions tend to make the maximization principles that govern simple competitive systems ineffective. I then review recent theoretical advances in the relationship between biodiversity and ecosystem functioning in food webs with multiple trophic levels. Last, I present a brief overview of a new theoretical approach that integrates nontrophic interactions in ecosystems while at the same time preserving the mass-balance constraints of ecosystem models and the dynamical flexibility of community models.

The synthesis presented in this chapter is still incomplete because it does not deal with complex network topologies as they are often observed in natural food webs (Cohen et al. 1990; Williams and Martinez 2000) and mutualistic networks (Bascompte et al. 2003). But it offers some principles and approaches that should help illuminate the functioning of these more complex systems. The issue of the stability of these systems will be examined specifically in the next chapter.

TROPHIC INTERACTIONS, VERTICAL DIVERSITY, AND ECOSYSTEM FUNCTIONING

The *trophic cascade* concept has traditionally been used to predict and interpret patterns of biomass at different trophic levels as functions of the presence of higher trophic levels and the fertility of the environment (Hairston et al. 1960; Oksanen et al. 1981; Carpenter et al. 1985). The same approach, however, can be used to predict the effects of *vertical diversity*—defined here as the number of trophic levels in an ecosystem—and environmental fertility on ecosystem properties such as the biomass, productivity, and ecological efficiency of the various trophic levels. In this section I deliberately reduce the enormous complexity of food webs and represent them as linear food chains with discrete trophic levels with a view to exploring the functional consequences of vertical diversity in its simplest form. As we shall see, the classical food chain is far from having given up all its secrets yet. Here I derive a number of new predictions that the theory of trophic cascades has not considered traditionally.

Consider a food chain of arbitrary length n, in which each trophic level i obeys a dynamical equation based on a mass or energy budget as in chapter 1 (figure 4.1). Since most ecosystems are thought to be limited by nitrogen or phosphorus, I shall use the mass of the limiting nutrient as my unit of measurement, but the same approach could be used for energy or any other nutrient as long as nutrient recycling is ignored and the energetic

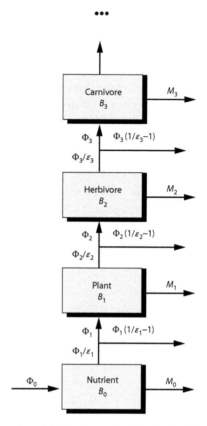

FIGURE 4.1. General model of a linear food chain. Each trophic level has a biomass B_i, a production Φ_i, a production efficiency ε_i, a loss flux M_i that includes basal metabolism and mortality, and a consumption flux to the next higher trophic level equal to the latter's production divided by its production efficiency. Boxes represent nutrient stocks, while arrows represent nutrient flows.

content and stoichiometric composition of the various trophic levels are roughly equal. Equation (1.19) is easily generalized in terms of the basic ecosystem processes of consumption and production at the various trophic levels as follows:

$$\frac{dB_i}{dt} = \Phi_i - M_i - \Phi_{i+1}/\varepsilon_{i+1}, \qquad 0 \le i \le n. \tag{4.1}$$

In this equation, B_i is the biomass of trophic level i, Φ_i is its production, ε_i is its production efficiency (as defined in chapter 1), M_i is a loss flux that includes basal metabolism and mortality, and the last term represents consumption by the next higher trophic level and is equal to the latter's production divided by its production efficiency (except, of course, for the last trophic level: $\Phi_{n+1} = 0$). Trophic level 0 is simply the inorganic nutrient, for which I assume that there is a constant input; i.e., $\Phi_0 = I$. This constant input incorporates both endogenous and exogenous sources of inorganic nutrient; at this stage, I am not considering nutrient cycling explicitly. The ecological efficiency of trophic level i, λ_i, is further defined as the ratio of its production to the production of the next lower trophic level (Lindeman 1942):

$$\lambda_i = \Phi_i/\Phi_{i-1}, \qquad 1 \leq i \leq n. \tag{4.2}$$

Finally, I assume, as in the classical theory of exploitation interactions, that there is no interference among consumers, so that the production and loss fluxes can be written as

$$\Phi_i = f_i(B_{i-1})B_i, \qquad 1 \leq i \leq n, \tag{4.3}$$

$$M_i = m_iB_i, \qquad 0 \leq i \leq n, \tag{4.4}$$

where $f_i(B_{i-1})$ is the functional response of trophic level i scaled by ε_i, and m_i is its mass-specific loss rate.

The first question that we can study using this simple, fairly general food-chain ecosystem model is: how does vertical diversity affect ecosystem functioning? In particular, does the addition of higher trophic levels enhance this functioning, does it make the ecosystem more efficient in some way, and, if so, in which way? To answer these questions, assume that the ecosystem reaches a stable equilibrium[1] and that environmental fertility, as measured by nutrient input I, is constant. The effects of vertical diversity on ecosystem properties can then be studied by comparing the biomass, production, and ecological efficiency of the various trophic levels at equilibrium as the number of trophic levels varies. The equilibrium values of biomass and ecological efficiency are provided in table 4.1 for ecosystems that range from 0 (only the inorganic nutrient is present) to 3 (the inorganic nutrient, plants, herbivores, and carnivores are present) trophic levels. In this table and in what follows, $X^*_{i(n)}$ denotes the equilibrium value of variable X at trophic level i in a system with n trophic levels. Note that the production of each trophic level i can be obtained simply by using the following formula derived from equation (4.2):

$$\Phi^*_{i(n)} = I\prod_{j=1}^{i}\lambda^*_{j(n)}. \tag{4.5}$$

---

[1] It is easy to show that the equilibrium is always stable for small food chains with $n < 2$ (i.e., food chains that contain only an inorganic nutrient and plants) and that it is stable for at least some parameter values for longer food chains with $n \geq 2$, depending on the form of the consumer functional responses.

TABLE 4.1. Equilibrium Values of the Biomass (B) and Ecological Efficiency (λ) of the Various Trophic Levels as the Number of Trophic Levels n Varies from 0 to 3 in the Model Ecosystem Described by Equations (4.1)–(4.4)[a]

	$n = 0$	$n = 1$	$n = 2$	$n = 3$
Biomass				
$B^*_{0(n)}$	I/m_0	$f_1^{-1}(m_1)$	$g_2^{-1}(B^*_{1(2)})$	$g_3^{-1}(B^*_{1(3)})$
$B^*_{1(n)}$		$\dfrac{\varepsilon_1}{m_1}(I - m_0 B^*_{0(1)})$	$f_2^{-1}(m_2)$	No explicit solution
$B^*_{2(n)}$			$\dfrac{\varepsilon_2}{m_2}[\varepsilon_1(I - m_0 B^*_{0(2)}) - m_1 B^*_{1(2)}]$	$f_3^{-1}(m_3)$
$B^*_{3(n)}$				$\dfrac{\varepsilon_3}{m_3}\{\varepsilon_2[\varepsilon_1(I - m_0 B^*_{0(3)}) - m_1 B^*_{1(3)}] - m_2 B^*_{2(3)}\}$
Ecological efficiency				
$\lambda^*_{1(n)}$		$\varepsilon_1\!\left(1 - \dfrac{B^*_{0(1)}}{B^*_{0(0)}}\right)$	$\varepsilon_1\!\left(1 - \dfrac{B^*_{0(2)}}{B^*_{0(0)}}\right)$	$\varepsilon_1\!\left(1 - \dfrac{B^*_{0(3)}}{B^*_{0(0)}}\right)$
$\lambda^*_{2(n)}$			$\varepsilon_2\!\left[1 - \dfrac{m_1}{f_1(B^*_{0(2)})}\right]$	$\varepsilon_2\!\left[1 - \dfrac{m_1}{f_1(B^*_{0(3)})}\right]$
$\lambda^*_{3(n)}$				$\varepsilon_3\!\left[1 - \dfrac{m_2}{f_2(B^*_{1(3)})}\right]$

[a] $X^*_{i(n)}$ denotes the equilibrium value of variable X at trophic level i in a system with n trophic levels, f_i^{-1} and g_n^{-1} denote the inverse functions of f_i and g_n, and function g_n is defined by the equation $B^*_{1(n)} = g_n(B^*_{0(n)}) = \varepsilon_1(I - m_0 B^*_{0(n)})/f_1(B^*_{0(n)})$.

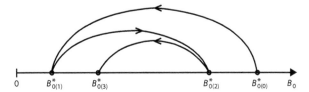

$$
0 \quad B^*_{0(1)} \qquad B^*_{0(3)} \qquad\qquad\qquad B^*_{0(2)} \qquad B^*_{0(0)} \qquad B_0
$$

FIGURE 4.2. Effects of increasing food-chain length, n, on the equilibrium amount of inorganic nutrient, $B^*_{0(n)}$, in the linear food chain depicted in figure 4.1. The amount of inorganic nutrient, just like the biomass of any trophic level, is highest when it lies at the top of the food chain, is lowest when it lies just below the top of the food chain, and jumps from low to high and high to low values but converges on an intermediate value as more trophic levels are added.

Despite the fact that not all of these equilibrium values have an explicit solution and some of them are fairly complicated, it is possible to determine how they change qualitatively as more trophic levels are added to the food chain (appendix 4A). The equilibrium biomasses of the various trophic levels satisfy the following inequalities:

$$B^*_{0(1)} < B^*_{0(3)} < B^*_{0(2)} < B^*_{0(0)}, \tag{4.6}$$

$$B^*_{1(2)} < B^*_{1(3)} < B^*_{1(1)}, \tag{4.7}$$

$$B^*_{2(3)} < B^*_{2(2)}. \tag{4.8}$$

These inequalities show a striking general pattern: the biomass of any trophic level is highest when it lies at the top of the food chain (there are no trophic levels above it), is lowest when it lies just below the top of the food chain (there is one trophic level above it), and jumps from low to high and high to low values but converges on an intermediate value as more trophic levels are added (figure 4.2).

Primary production and ecological efficiencies show the same pattern, although quantitatively they vary much less than biomasses:

$$\Phi^*_{1(2)} < \Phi^*_{1(3)} < \Phi^*_{1(1)}, \tag{4.9}$$

$$\lambda^*_{1(2)} < \lambda^*_{1(3)} < \lambda^*_{1(1)}, \tag{4.10}$$

$$\lambda^*_{2(3)} < \lambda^*_{2(2)}. \tag{4.11}$$

Last, production at the second (herbivore) trophic level can either increase or decrease as the third (carnivore) trophic level is added, depending on the form of the functional response of the first (plant) trophic level.

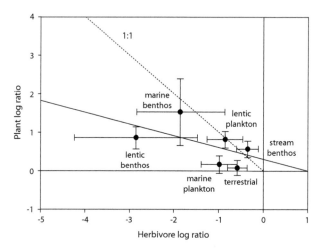

FIGURE 4.3. The effect size (log ratio) of predators on primary producers vs. herbivores in six major ecosystem types. Error bars are 95 percent confidence intervals. The effect of predators is significant if the confidence interval does not overlap zero. The solid line shows the linear regression relating the plant and herbivore effect sizes, while the dotted line shows the 1:1 relationship. Modified from Shurin et al. (2002).

Several general conclusions emerge from this theoretical reexamination of the classical food chain:

1. *The top-down control exerted by the top trophic level cascades down the food chain but becomes progressively weaker*, so that lower trophic levels are less and less affected by the addition of further trophic levels at the top. The biomass of a trophic level varies from high values when this trophic level lies at the top of the food chain, or at an even number of levels below it, to low values when it lies at an odd number of levels below the top, but these oscillations in biomass dampen as more trophic levels are added higher in the food chain. This predicted attenuation of the trophic cascade down the food chain is consistent with experimental data. Experiments that have manipulated carnivore presence show that carnivores generally have a stronger effect on herbivore biomass than on plant biomass (McQueen et al. 1986; Shurin et al. 2002) (figure 4.3). Note that my theoretical predictions consider the effects of adding a variable number of trophic levels on the biomass of a single target trophic level, whereas experiments have considered the effects of adding a single trophic level on the biomass of several lower trophic levels.

Despite this difference, experimental data are at least qualitatively consistent with theoretical predictions.

2. The same cascading but attenuated top-down effects along the food chain hold for ecological efficiencies, but these are much less affected quantitatively than are biomasses. Production is even less affected, and secondary production can even sometimes increase upon addition of a third trophic level. Thus, as a rule, *production should be little affected by trophic cascades*. Unfortunately, I know of no empirical studies that have tested trophic cascades for production.

3. The final, disturbing conclusion is that *vertical diversity does not maximize any ecosystem property*. In contrast to horizontal diversity, which tends to enhance biomass and production in simple competitive systems through functional complementarity among species, vertical diversity tends to bring ecosystems toward a sort of medium functioning. Both the biomasses and ecological efficiencies of the various trophic levels tend to converge to intermediate values as food-chain length increases.

The latter conclusion, however, holds for ecosystem properties measured at the scale of specific trophic levels. What about some more integrative measures of the functioning of the whole ecosystem, such as total ecosystem biomass or total ecosystem production cumulated over all trophic levels? I show in appendix 4A that total ecosystem biomass should on average stay constant as food-chain length increases when the production efficiencies of the various trophic levels are maximal ($\varepsilon_i = 1$), and they should decrease under the more realistic conditions where production efficiencies are less than maximal ($\varepsilon_i < 1$). This conclusion makes sense since materials and energy are gradually lost along the food chain when there is no internal recycling as in the model considered here. By contrast, total cumulative ecosystem production is expected to increase slightly overall as food-chain length increases, although variations in primary production due to changes in top-down control may override the production increments of the additional trophic levels. We shall see in chapter 6 that nutrient cycling has a much stronger potential to enhance production in ecosystems—not only production cumulated over different trophic levels, which has questionable relevance for both basic and applied purposes, but also production at each trophic level. My purpose here is to explore the functional consequences of vertical diversity per se, and they appear not to be overwhelmingly positive for any measure of ecosystem functioning, except to some extent for total

TABLE 4.2. Direction of the Changes in the Biomass (B), Production (Φ), and Ecological Efficiency (λ) of the Various Trophic Levels at Equilibrium as Environmental Fertility I Increases.[a]

	$n = 0$	$n = 1$	$n = 2$	$n = 3$
Biomass				
$B^*_{0(n)}$	+	0	+	$- (0\,+)$
$B^*_{1(n)}$		+	0	+
$B^*_{2(n)}$			+	0
$B^*_{3(n)}$				+
Production				
$\Phi^*_{1(n)}$		+	+	+
$\Phi^*_{2(n)}$			+	+
$\Phi^*_{3(n)}$				+
Ecological efficiency				
$\lambda^*_{1(n)}$		+	$- (0\,+)$	+
$\lambda^*_{2(n)}$			+	$- (0\,+)$
$\lambda^*_{3(n)}$				+

[a] $X^*_{i(n)}$ denotes the equilibrium value of variable X at trophic level i in a system with n trophic levels. When several signs are present, the outcome is indeterminate but the signs in parentheses are less probable.

cumulative ecosystem production. We shall continue to explore this intriguing conclusion throughout this chapter.

The second question that we can address using our simple food-chain ecosystem model is: how does *environmental fertility* affect ecosystem functioning for a given number of trophic levels? The classical trophic cascade theory based on logistic plant growth and type-2 consumer functional responses (Oksanen et al. 1981) predicts that the biomass of the trophic levels that lie at the top of the food chain or at an even number of levels below it will respond positively to fertilization, whereas the other trophic levels will not respond to fertilization because they are top-down-controlled. But we may wish to be slightly more general and see if this pattern holds for other functional responses, as well as for production and ecological efficiency. This question can be studied using the derivative of the equilibrium values of the biomass, production, and ecological efficiency of the various trophic levels provided in table 4.1 with respect to parameter I, which measures environmental fertility. The results are presented in appendix 4B and summarized in table 4.2.

Table 4.2 shows that the predictions of the classical trophic cascade theory regarding biomass hold generally, at least for the top three trophic levels. Note that the stock of inorganic nutrient departs from these predictions since it is expected to often decrease, and sometimes increase, upon nutrient enrichment in the presence of carnivores. An examination of a food chain with four levels would confirm this trend since the responses of the equilibrium stocks of both the inorganic nutrient and primary producers then become indeterminate. These specific results highlight the advantage of studying genuine ecosystem models in which the inorganic nutrient is represented explicitly. Previous models did not consider inorganic nutrients explicitly (Oksanen et al. 1981). In contrast to biomass, production at all levels always increases with fertilization. This emphasizes once more the difference between biomass and production, between stocks and fluxes. As a rule, *production is less affected by top-down forces than is biomass or population density* and responds more to bottom-up influences because of simple mass-balance constraints: an increased inflow at the bottom must necessarily be balanced by an increased outflow in the long run. Part of this outflow is the inflow of the next trophic level, which in turn has an increased outflow, and so on up to the top of the food chain, so that the effect is propagated to the entire system. Despite increased productivity, however, the ecological efficiency of the trophic level just below the top is expected to generally decrease (table 4.2). This occurs because the biomass of this trophic level is top-down-controlled, and hence its production responds relatively less than that of the next lower trophic level.

DOES EVOLUTION OR SPECIES TURNOVER ENHANCE ECOSYSTEM FUNCTIONING IN FOOD WEBS?

We have seen in the previous section that vertical diversity tends to bring ecosystem processes and properties within trophic levels toward some medium, rather than maximum, level in a constant environment. But might this leveling off of ecosystem properties be compensated for by evolution toward more efficient types or species replacement by more efficient species in food webs? Bob Holt and I have shown that species replacement at one trophic level does not necessarily lead to enhanced ecosystem functioning at all levels (Holt and Loreau 2001). There is a fundamental reason why we should expect evolution and species turnover to enhance ecosystem processes only up to a certain point, after which they tend to maintain them at a suboptimal level or even to make them deteriorate.

Evolution through natural selection at one hierarchical level does not occur for the good of all and may even have negative consequences for the next higher hierarchical level. It is well known that classical Darwinian evolution at the individual level does not necessarily benefit the population as a whole. The same conclusion applies even more to communities and ecosystems, in which the potential for counterintuitive feedbacks and indirect interactions is greater. I shall take a simple classical example to illustrate this point using one of the simplest possible models of a trophic interaction between a prey and its predator, proposed by Rosenzweig and MacArthur (Rosenzweig and MacArthur 1963; Rosenzweig 1971, 1973).

Ignore mass-balance constraints for the time being and assume that the prey, with population size or biomass N, has a population growth that obeys the logistic equation and that the predator, with population size or biomass P, has a type-2 functional response:

$$\frac{dN}{dt} = rN\left(1 - \frac{N}{K}\right) - \frac{cNP}{D + N},$$
$$\frac{dP}{dt} = \frac{\varepsilon cNP}{D + N} - mP. \tag{4.12}$$

In these equations, r is the intrinsic rate of natural increase of the prey, K is its carrying capacity, c is the maximum consumption rate of the predator, D is its half-saturation constant, ε is its production efficiency, and m is its mortality rate.

The asymptotic behavior of this prey–predator system can be studied graphically by an isocline analysis (figure 4.4). The null isoclines of the prey and predator are, respectively,

$$P = \frac{r}{cK}[DK + (K - D)N - N^2] \tag{4.13}$$
$$N = \frac{mD}{\varepsilon c - m}.$$

The predator isocline is vertical and sets the equilibrium value of the prey population size because of the top-down control exerted by the predator on the prey. The prey isocline is hump-shaped when $K > D$ (the linear term is then positive and is greater than the negative quadratic term for small N) and monotonic decreasing otherwise. It is well known that the equilibrium that lies at the intersection of the two isoclines is stable when it lies in the right, descending part of the prey isocline, whereas it is unstable and gives rise to a stable limit cycle when it lies in the left, ascending part of the prey isocline (May 1973).

Now, assume that evolution occurs within the predator population by natural selection. Individual predators that have the highest net growth rate will be selected for, and a higher net growth rate can be achieved by either increasing parameter ε or c or decreasing parameter D or m. All these

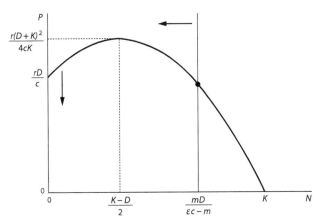

FIGURE 4.4. Effects of predator evolution on the equilibrium population sizes of prey and predators in the Rosenzweig-MacArthur model. The ecological equilibrium point lies at the intersection of the hump-shaped prey null isocline and the vertical predator null isoclines. Predator evolution shifts the predator isocline to the left, and the left part of the prey isocline downward. As a result, the equilibrium prey population size decreases, while the equilibrium predator population size first increases and then decreases.

changes will have the effect of shifting the predator isocline to the left: the predator becomes more efficient at exploiting the prey and hence depresses the prey population further. If natural selection acts on parameter c or D that determines prey consumption by the predator, the prey isocline will also be affected, such that its maximum shifts to the right and its y-intercept declines (figure 4.4). All these changes have two obvious effects on the prey–predator interaction: (1) they decrease the prey population size, and (2) they tend to destabilize the system since the equilibrium can become unstable if it is brought to the left of the hump of the prey isocline.

Their effect on the predator population is slightly more complex. If we start from a situation in which the predator is relatively inefficient, such that its isocline lies far on the right of the hump of the prey isocline, evolution will gradually shift its isocline to the left, which at first increases its population size (the equilibrium point at the intersection of the two isoclines moves up). If, however, evolution toward greater efficiency continues further, the predator isocline will eventually move past the hump (which itself may shift to the right), at which point the predator population starts decreasing. Since the y-intercept of the prey isocline moves downward simultaneously, both the predator and prey populations ultimately tend to zero. In fact, this point will never be reached since the system will become increasingly unstable, generating fluctuations of increasing amplitude, as the predator becomes more efficient at exploiting the prey. Thus, stochastic extinction will occur before the deterministic extinction of the two interacting populations can ever take place.

The general conclusion, however, is clear: evolution toward more efficient predators at first benefits the predator population, but, past a threshold, it decreases both the prey and predator populations, ultimately driving them to extinction. Species turnover, in which more efficient predators replace less efficient ones by successive invasions from outside the ecosystem, leads to exactly the same result. It is hard to find a clearer example of how *evolution in food webs can be detrimental to the functioning, and even the very existence, of the ecosystem*. Of course, the prey too will evolve in response to the deterioration of its selective environment, leading to coevolution of the two partners and possibly stabilization of the interaction. But there is no reason to expect this coevolution to lead to optimal functioning, whatever the criterion one wishes to choose to define optimal functioning. Darwinian extinction, i.e., the process by which individual natural selection leads to population extinction, might be a widespread, if often ignored, phenomenon. Webb (2003) provides a classification of the various dynamical mechanisms that can generate Darwinian extinction.

This simple example reveals a general property of trophic interactions: the potential for *overexploitation*. Overexploitation occurs when increasing the exploitation of a resource leads to a decreased yield to the consumer. Overexploitation does not occur (at least at equilibrium) when resources are inorganic or inert because such resources do not self-reproduce but are renewed by independent factors (whether through recycling within the ecosystem or through inputs from outside the ecosystem). Increasing the exploitation rate of these resources does not affect their renewal and hence can only increase consumption up to a level where it matches resource renewal. As a result, overexploitation of inorganic resources by primary producers is unlikely, except as a transient phenomenon. In contrast, living resources have the fundamental property of being self-reproducing. Exploiting them beyond a certain threshold reduces their population size to such an extent that their collective production, and hence also the production available for their exploitation, decreases, leading ultimately to the decline of the exploiter populations. Some of the best-known examples of overexploitation involve our own species since humans have driven a number of large-sized mammals to extinction or near extinction because of hunting, and a similar fate is threatening a growing number of large-sized fishes because of increasing fishing efforts. Infectious disease ecology is another area where overexploitation is widespread and conspicuous. Infectious diseases often decimate host populations and subsequently fade out for lack of susceptible hosts. The commonness of overexploitation in many other natural systems, however, is still poorly known.

HORIZONTAL DIVERSITY, VERTICAL DIVERSITY, AND ECOSYSTEM FUNCTIONING

The previous chapters showed that horizontal diversity tends to enhance resource exploitation and hence also generally production and biomass, within a trophic level. Therefore, we might intuitively expect horizontal diversity to counteract the disruptive effect of vertical diversity on ecosystem functioning. This should be true in particular at the bottom of the food web. If inorganic resources are not fully exploited by primary producers because of the pressure exerted by higher trophic levels, other species should be able to invade the system and use the resource leftovers, thus restoring a higher primary production and biomass. As it turns out, things are not so simple because there are strong constraints on coexistence in exploitation systems.

ASSEMBLY RULES AND ECOSYSTEM FUNCTIONING IN A FOOD WEB WITH PLANTS AND SPECIALIST HERBIVORES

The constraints that arise from species interactions in exploitation systems are particularly clear in the case of specialist plant–herbivore food chains supported by a single, homogeneous limiting nutrient. Grover (1994) showed that strict rules govern community assembly in this case. In his analysis, he made some restrictive assumptions about mortality rates (which were all equal to a common dilution rate, as in a chemostat) and nutrient recycling, but his results are robust to relaxation of these assumptions. To understand the functional consequences of these constraints on coexistence, consider the general food-chain model described in the first section but assume that there is an arbitrary number of such food chains, each of which is limited to two trophic levels, i.e., plants and herbivores (figure 4.5). The dynamical equations of the system depicted in figure 4.5 are

$$\frac{dB_0}{dt} = I - m_0 B_0 - \sum_j f_{1j}(B_0) B_{1j}/\varepsilon_{1j},$$

$$\frac{dB_{1i}}{dt} = f_{1i}(B_0) B_{1i} - m_{1i} B_{1i} - f_{2i}(B_{1i}) B_{2i}/\varepsilon_{2i}, \qquad (4.14)$$

$$\frac{dB_{2i}}{dt} = f_{2i}(B_{1i}) B_{2i} - m_{2i} B_{2i},$$

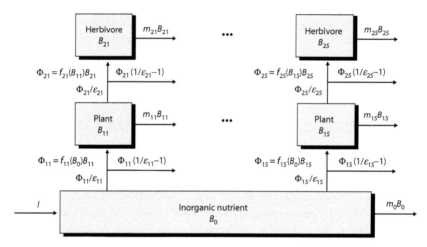

FIGURE 4.5. A simplified nutrient-limited food web made up of S specialist plant–herbivore food chains. Boxes represent nutrient stocks, while arrows represent nutrient flows.

where variables and parameters have two subscripts, the first of which indicates the trophic level, and the second, species identity.

In this system, the stock of inorganic nutrient obeys strict constraints at equilibrium (appendix 4C). Let $B_{0(\cdot)}^*$ denote the equilibrium value of the stock of inorganic nutrient in the presence of a community with composition $(\cdot)$ and write this composition using the subscripts of the species present[2]. Figure 4.6 illustrates the multiple inequalities that must be satisfied when there are up to two plant and herbivore species. It is straightforward to generalize these inequalities to an arbitrary number of food chains.

Since there is a single limiting resource in this system, competition among plants obeys the R^* rule in the absence of herbivores (chapter 2): the plant species with the lowest B_0^* drives all other species to extinction. Here I have arbitrarily defined species 1 as the most competitive one $(B_{0(0,11)}^* < B_{0(0,12)}^*)$. The addition of herbivores to a food chain increases B_0^* because the loss of plant biomass to herbivores must be compensated for by a higher plant growth rate at equilibrium, which itself requires a higher resource availability. Consequently, plant 1 can invade and extirpate all food-web configurations from which it is absent since these are characterized by a nutrient availability that exceeds its B_0^*. Plant 2 cannot invade a

---

[2] Thus, for instance, $B_{0(0,11,21)}^*$ is the equilibrium inorganic nutrient stock (first subscript 0) in a simple food chain that comprises the inorganic nutrient (subscript 0 in parentheses), plant species 1 (subscript 11 in parentheses), and herbivore species 1 (subscript 21 in parentheses).

system in which only plant 1 is present, but it can invade a system with a food chain made up of plant 1 and herbivore 1 provided its B_0^* is lower than that of food chain 1. This invasion, however, does not lead to the competitive exclusion of plant 1, and hence of herbivore 1, because plant 1 is a better competitor. Thus, as emphasized by Grover (1994), the constraints shown in figure 4.6 not only determine the potential for *species coexistence* but also the precise *assembly sequence* in which the community can assemble itself through successive species introductions. The unique sequence of introductions that leads to a community of two plants and their specialist herbivores limited by a single nutrient is the following: (1) plant 1 (2) herbivore 1 (3) plant 2 (4) herbivore 2. To be feasible, this sequence requires a trade-off between the ability of plants to compete for resource exploitation and their ability to resist herbivory, such that stronger competitors are most suppressed by their herbivores, thereby leaving enough nutrient available to weaker competitors.

During this community assembly process, the level of the basal inorganic resource alternatively decreases and increases at each introduction but converges on an intermediate value, a pattern that is strikingly similar to the assembly of a single food chain (figure 4.2). The inorganic nutrient stock in turn determines total resource consumption by plants: at equilibrium, the first of equations (4.14) implies that inorganic nutrient consumption (the summation term on the right-hand side of the equation) equals $I - m_0 B_0^*$. Therefore, inorganic nutrient consumption is lower when B_0^* is higher, and vice versa. And since primary production and plant biomass depend directly on inorganic nutrient consumption, unless there are special trade-offs among plant traits, they should be expected to generally show a pattern symmetrical to that of the inorganic nutrient stock (figure 4.6). Thus, we are led to the conclusion that *horizontal diversity has qualitatively the same effect on ecosystem properties as does vertical diversity* in food webs made up of plants and specialized herbivores. Each new plant species added to such food webs has qualitatively the same effect as does an additional consumer trophic level in a linear food chain, and each new herbivore species has qualitatively the same effect as does a second additional trophic level that controls the first.

The ups and downs of nutrient availability correspond to different foodweb configurations, however. Therefore, it is also useful to analyze the effects of increasing species diversity in the two configurations separately. When the food web includes a plant that is not controlled by its specialist herbivore (figure 4.6, left), simultaneously increasing plant and herbivore diversity along the assembly sequence leads to an increased available nutrient

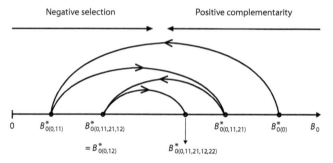

FIGURE 4.6. Coexistence conditions and assembly rules for the nutrient-limited food web with plants and specialist herbivores depicted in Figure 4.5. The unique sequence of introductions that leads to an equilibrium community of two plants and their specialist herbivores is the following: (1) plant 1; (2) herbivore 1; (3) plant 2; (4) herbivore 2. The amount of inorganic nutrient alternatively decreases and increases at each introduction but converges on an intermediate value, just as in a single food chain (figure 4.2). Adding a plant–herbivore pair to a system that contains an unconsumed plant generates a negative selection effect that deteriorates nutrient exploitation (left), while adding a plant–herbivore pair to a system in which all plants are controlled by their specialist herbivore generates a positive complementarity effect that enhances nutrient exploitation (right).

stock, and hence to a less efficient nutrient exploitation. In contrast, when all plants are controlled by their specialist herbivores (figure 4.6, right), simultaneously increasing plant and herbivore diversity leads to a decreased available nutrient stock and hence to a better nutrient exploitation. Thus, the effects of species diversity on nutrient consumption, and hence also generally on plant biomass and primary production, are expected to be opposite in the two alternating parts of the assembly process, and these contrasting responses are entirely due to differences in the configuration of the food web in the two parts. Plant–herbivore food chains are less efficient at using the inorganic nutrient than are plants alone. Adding a plant–herbivore pair to a system that contains an unconsumed plant can only be achieved, while maintaining equilibrium coexistence, if the herbivore consumes the most competitive plant, which counteracts selection toward a greater competitive ability in plants and reduces the ability of the system as a whole to exploit the limiting nutrient. On the other hand, if the system contains only plant–herbivore food chains, adding another food chain—as long as the inorganic nutrient input is large enough to support it—does not affect the plants that are already present because their biomass is top-down-controlled by the herbivores. Thus, increasing the diversity of full plant–herbivore food chains can only increase the plants' total biomass and collective efficiency at exploiting the inorganic nutrient.

This contrast between the two food-web configurations shows that the two main biodiversity effects at the plant trophic level, the selection effect and the complementarity effect, are driven by different factors and may conflict when plant coexistence is mediated by a consumer trophic level. The *selection effect* in plants is driven by competition for a single limiting resource, which tends to maximize resource consumption. In contrast, the *complementarity effect* cannot arise from niche differences among plants in the way they use resources since the model assumes that there is a single, homogeneous limiting resource without any means of partitioning it. Here, complementarity among plants arises from *avoidance of herbivore-mediated or "apparent" competition* (Holt 1977) because they have different specialized consumers. This new form of functional complementarity that arises from trophic interactions has a positive effect on plant ecosystem processes when all plants are involved in trophic interactions. But plants can always gain greater access to the shared limiting resource if they escape trophic interactions with the upper trophic level. Therefore, this form of complementarity is unable to compensate for the intrinsically detrimental effect that trophic interactions have on plant ecosystem processes, and it conflicts with selection toward greater resource acquisition among those plants that escape trophic interactions and compete for the limiting resource.

These conclusions are important because they provide insights into some basic biodiversity effects that operate in food webs and their mechanisms. But they do not necessarily apply to all natural ecosystems. Model (4.14) makes, implicitly or explicitly, strong assumptions about the nature of species interactions and the assembly process. By assuming a single, homogeneous limiting nutrient, the model ignores any form of functional complementarity that arises from resource partitioning or facilitation among plants, for which there is now ample theoretical and experimental support as shown in chapter 3. By assuming that the system reaches equilibrium at each step of the assembly sequence, the above analysis ignores the possibility that new species may invade during transient dynamics, and hence that biodiversity loss or gain may not follow a neat assembly sequence.

BIODIVERSITY AND ECOSYSTEM FUNCTIONING IN FOOD WEBS

Elisa Thébault and I (Thébault and Loreau 2003, 2006; Loreau and Thébault 2005) developed a model that partly relaxes some of the restrictive assumptions involved in Grover's (1994) assembly rules and that is more suitable for a general analysis of the relationship between biodiversity and ecosystem functioning in food webs. First, our model considers consumers whose dietary niche breadth can vary from specialist to generalist and thereby relaxes the assumption that each plant species is controlled by a unique specialist herbivore. It also includes carnivores as a third trophic level, although most of its interesting features can be analyzed with only two trophic levels. Second, the model allows some degree of coexistence and resource partitioning among plants through limited access to the limiting nutrient in local resource depletion zones around the rooting system of each plant, following the formalism developed by Huston and DeAngelis (1994) and me (Loreau 1996, 1998a). Our analysis, however, focused mainly on the case where nutrient transport in the soil is relatively fast, and hence where the potential for plant coexistence through this mechanism is limited, to emphasize the specific role trophic interactions play in the relationship between biodiversity and ecosystem functioning. Last, our analysis of the model considered the expected response of ecosystem properties following random species loss or gain to mimic recent biodiversity experiments and study the effects of species diversity independently of species identity. Consequently, we relaxed the assumption that changes in species richness can occur only as a result of a strict, sequential community assembly process since this strong assumption is irrelevant in the case of biodiversity loss and community disassembly.

Our model is depicted graphically in figure 4.7. The corresponding dynamical equations are obtained easily by setting the rate of change of each compartment equal to the sum of inflows to that compartment minus the sum of outflows from that compartment. Here, R is the nutrient stock in the soil nutrient pool with volume V_R, L_i is the nutrient stock in the set of individual resource depletion zones, (with total volume V_i) of plants from species i, and P_i, H_i, and C_i are the nutrient stocks of plant, herbivore, and carnivore species i, respectively. Biomasses are again assumed to be simply proportional to nutrient stocks here. $\sigma_i = V_i/V_R$ is the relative volume occupied by plant species i in the soil[3]. The limiting nutrient is supplied in inorganic form with an amount I per unit time, is lost at a rate q per unit time, and is transported between individual resource depletion zones and

---

[3] The variables in figure 4.7 are nutrient stocks in order to make mass balance explicit and obtain direct measures of stocks and biomasses. Earlier models (Loreau 1996, 1998) were expressed in terms of nutrient concentrations. The model described by figure 4.7 is equivalent to these earlier models after nutrient concentrations are multiplied by the appropriate volumes to obtain nutrient stocks. The original model presented in Thébault and Loreau (2003), however, had two types of variables: nutrient concentrations for the inorganic nutrient and nutrient stocks for the living compartments.

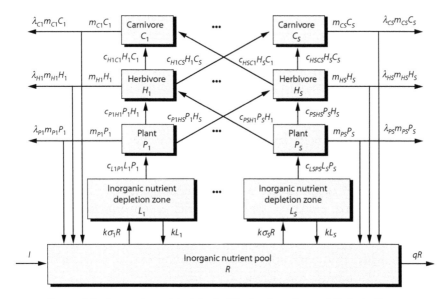

FIGURE 4.7. A complex nutrient-limited food web made up of S plants, S generalist herbivores, and S generalist carnivores. Plants have limited access to inorganic nutrient in individual resource depletion zones. Boxes represent nutrient stocks, while arrows represent nutrient flows.

the soil nutrient pool at a rate k per unit time. Plant species i consumes the inorganic nutrient at a rate c_{LiPi}, herbivore species j consumes plant species i at a rate c_{PiHj}, and carnivore species j consumes herbivore species i at a rate c_{HiCj}. The nutrient is released from biomass by natural death and metabolic activity at rates m_{Pi}, m_{Hi}, and m_{Ci} in plants, herbivores, and carnivores, respectively. Part of the nutrient thus released is recycled within the local ecosystem; λ_{Pi}, λ_{Hi}, and λ_{Ci} represent the fractions of nutrient coming from plants, herbivores and carnivores, respectively, that are not recycled and hence are lost from the system. Resource–consumer interactions are here assumed to follow the law of mass action (i.e., consumer functional responses are assumed to be linear), and production efficiencies are assumed to be 100 percent for the sake of simplicity. Lower production efficiencies do not change the results qualitatively; they affect only the shape of the nutrient or biomass pyramid of the ecosystem, i.e., the distribution of nutrient across the various trophic levels. The number of species is in principle S per trophic level, but it may differ between trophic levels in some food-web configurations.

The main effect of carnivores in this model is to control the herbivores they prey upon and thereby to release plants from the top-down control of herbivores. This creates a situation that is qualitatively similar to that in which some plants are unconsumed by herbivores. Therefore, I consider here only cases in which carnivores are absent for simplicity. I also present results only for total plant biomass, total herbivore biomass, and total ecosystem biomass (sum of total plant and herbivore biomasses) as ecosystem properties for comparison with experimental studies. Total plant biomass is likely to be a poor approximation of primary production when consumed by higher trophic levels, but total biomasses at producer and consumer trophic levels are often used for convenience as ecosystem properties in experiments on the relationship between species diversity and the functioning of both single-trophic-level and multitrophic ecosystems (Duffy et al. 2003; Finke and Denno 2005). In the simple scenarios analyzed here, all plants are assumed to have equal nutrient turnover rates m_{P_i}, in which case primary production can be shown to be proportional to total ecosystem biomass. Thus, total ecosystem biomass can be used as a convenient surrogate for primary production.

Using this model, we examined how changes in species richness influence ecosystem properties at equilibrium for different food-web structures under conditions that allow all plant and herbivore species to coexist. We also considered different scenarios of biodiversity changes: either plant species richness and herbivore species richness vary in parallel or herbivore species richness varies alone. Changing plant richness alone leads to unfeasible food-web configurations in our model because there cannot be more herbivore species than plant species at equilibrium. To analyze expected ecosystem responses to changes in species richness, we calculated, at each diversity level, the expected value of plant and herbivore biomass across all possible species compositions or, equivalently, in randomly assembled communities, as is often done in experiments.

The most striking conclusion of our analysis is that *the effects of species richness on ecosystem properties are critically dependent on the structure of the food web.* First, as expected from the previous analysis of specialist plant–herbivore food webs in a homogeneous environment, the *presence of plants that are released from top-down control by herbivores*, either because they are inedible or because their herbivores are themselves controlled by carnivores, strongly affects the relationships between diversity and biomass. When each plant is consumed by a specialist herbivore, the mean total plant biomass increases linearly with species richness (figure 4.8, left) because each plant species is controlled by its own herbivore and is unaffected by the addition of other species. The corresponding complementarity effect (as defined in chapter 3) is positive, while the selection effect is

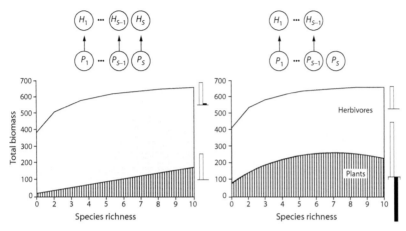

FIGURE 4.8. Expected total plant biomass (hatched area) and total herbivore biomass (gray area) as functions of plant species richness, for a food web with specialist herbivores and either no inedible plant (left panel) or one inedible plant (right panel) in the model depicted in figure 4.7 but without carnivores. The upper curve shows total ecosystem biomass (sum of total plant and herbivore biomasses). Total herbivore biomass is the difference between the upper and lower curves (gray area). Herbivore species richness varies parallel to plant species richness to keep the same food-web configuration along the diversity gradient. Top panels show the food-web configurations analyzed in the corresponding column. Small histograms on the right of the panels show the strengths of the complementarity effect (in white) and the selection effect (in black) for the highest diversity treatment (10 plant species). These effects are measured on the same scale as the y-axis, except for plant biomass in case of a food web with one inedible plant, where they are reduced by a factor of 2. Modified from Loreau and Thébault (2005).

zero. As discussed in the previous section, this complementarity is generated by a very different mechanism than in simple competitive systems: here it does not arise from resource partitioning or facilitation but from avoidance of herbivore-mediated competition through differentiation in plants' natural enemies.

In contrast, when the same food web comprises a plant that is either inedible or protected from top-down control by a carnivore, the mean total plant biomass does not increase linearly and can even decrease at high diversity (figure 4.8, right). In this case, the biomass of the inedible plant is controlled by resource availability, which decreases when plant richness increases. This also leads to a negative selection effect because the inedible plant, which tends to be dominant is most affected by an increase in diversity. Note that, in this scenario, since our analysis considers the expected biomass across all possible species compositions, some assemblages do not contain the inedible plant, which explains why the mean total plant biomass increases with diversity over at least part of the diversity gradient. If the inedible plant were included automatically in all assemblages, total plant biomass would decline monotonically with diversity, as predicted by the assembly rule analyzed in the previous section. This illustrates the fact that specific scenarios of biodiversity loss or gain may deviate significantly, and even qualitatively, from the expected response.

In both cases, total herbivore biomass (difference between the upper and lower curves in figure 4.8) can show complex relationships with diversity. In the scenarios examined in figure 4.8, it decreases at high diversity when all plants are edible, and it also decreases at intermediate species richness when the food web comprises an inedible plant. Total ecosystem biomass (upper curves in figure 4.8) and primary production, however, increase monotonically with diversity until saturation, just as in systems with a single trophic level. Thus, the *nature of population control* (top-down vs. bottom-up) in an ecosystem can profoundly affect the responses of ecosystem properties to changes in species richness. Heterogeneity within trophic levels and the presence of inedible species are important to consider as they modify top-down control and trophic cascades in food webs (Leibold 1989; Abrams 1993).

Food-web connectivity, as measured by the diet breadth of herbivores, is another factor that has a strong impact on the relationship between diversity and ecosystem properties (figure 4.9, top). When all plants are consumed and herbivores are generalists, the mean total plant biomass no longer increases linearly with diversity and can even decrease at high diversity levels. In this case, the biomass of each plant species is still controlled by herbivores, but it decreases with the addition of other herbivore species because plant consumption increases. This in turn can result in decreased total plant biomass. The mean total herbivore biomass is generally higher when herbivores are generalists than when they are specialists, but it also increases less with diversity and can decrease at high diversity. Competition between generalist herbivores is strong, and resource-use complementarity among them is lower, as indicated by the smaller complementarity effect. Total ecosystem biomass and primary production can also decrease at high diversity when herbivores are generalists.

The *trophic position* of the species being lost or gained also plays a critical role. Changes in the diversity of the consumer trophic level alone (figure 4.9, bottom) have very different effects than do simultaneous changes at the plant and herbivore trophic levels (figure 4.9, top). The mean total

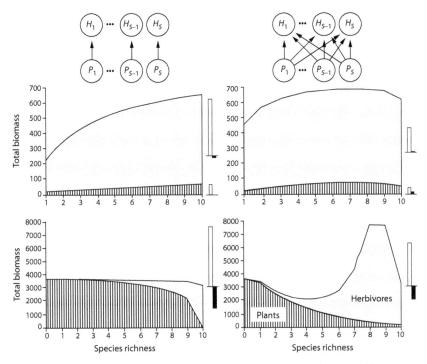

FIGURE 4.9. Expected total plant biomass (hatched area) and total herbivore biomass (gray area) as functions of species richness, for a food web with specialist herbivores (left panels) and a food web with generalist herbivores (right panels) in the model depicted in figure 4.7 but without carnivores. The upper curve in each panel shows total ecosystem biomass (sum of total plant and herbivore biomasses). Total herbivore biomass is the difference between the upper and lower curves (gray area). Species richness varies either at the two trophic levels simultaneously (top panels) or at the herbivore trophic level only while plant species richness is held constant at 10 species (bottom panels). Small histograms on the right of panels show the strengths of the complementarity effect (in white) and the selection effect (in black), measured on the same scale as the y-axis, for the highest diversity treatment (10 herbivore species). Modified from Loreau and Thébault (2005).

plant biomass then always decreases upon herbivore addition, whether herbivores are specialists or generalists. But it decreases faster at low diversity when herbivores are generalists because the consumption of each plant is then higher. The mean total herbivore biomass always increases with diversity when herbivores are specialists, but it can decrease at high diversity when herbivores are generalists (figure 4.9, bottom right). Again, when herbivores are generalists, resource-use complementarity is smaller, which can result in a decrease in total herbivore biomass at high diversity and a smaller complementarity effect. The strong increase in total herbivore biomass at an intermediate diversity of herbivores when they are generalists may be explained by a strong increase in herbivore consumption together with more favorable conditions for herbivore-mediated plant coexistence.

Using very different models of predator–prey interactions, Ives et al. (2005) and Casula et al. (2006) have further explored the effects of *nonadditive interactions* between consumers on the total density or biomass of both consumers (predators) and resources (prey). Nonadditive interactions arise when consumers either decrease (antagonism) or increase (synergism) the per capita capture rates of other consumer species because of nontrophic effects such as mutual interference (antagonism) and facilitation (synergism). Ives et al. (2005) included these interactions in the form of a single parameter a that modulates the per capita capture rate of consumers. If a is negative, increasing the density of a consumer species decreases the per capita capture rate of another consumer species, thus generating antagonism. The converse is true if a is positive, which leads to synergism. Their model shows patterns very similar to ours when consumers are generalists and only consumer species richness is varied (compare figure 4.9, bottom right, and figure 4.10, left panels): total resource density or biomass decreases, and total consumer density or biomass shows a hump-shaped pattern as initial consumer species richness increases. Synergistic interactions between consumers only strengthen these patterns, while antagonistic interactions weaken them and may sometimes turn a humped-shaped pattern of consumer density into a monotonic increase (figure 4.10, bottom left). This occurs because antagonistic interactions inhibit the extinction of resource species in their model (figure 4.10, top right), thereby weakening the negative effect of consumers on resources at high consumer richness. Resource extinction, however, is not required to generate such hump-shaped patterns. These also emerge in our model in spite of the fact that we considered only food webs in which there was no species extinction. Declines in consumer biomass or density at high consumer diversity arise from resource overexploitation, which is a more general phenomenon than outright extinction. Last, intraguild predation has effects very similar to those of antagonistic interactions because it also acts to reduce the impact of consumers on resources (Ives et al. 2005).

In conclusion, horizontal diversity does not simply oppose the effect of vertical diversity on ecosystem functioning. The interactions between horizontal diversity and vertical diversity are complex and lead to complex relationships between biodiversity and ecosystem properties. Horizontal

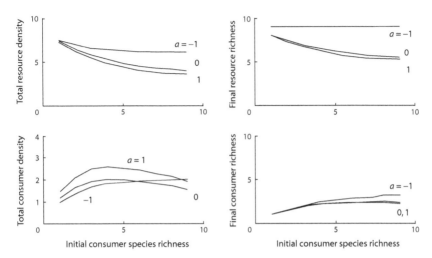

FIGURE 4.10. Effects of nonadditive interactions on the relationships between total resource density (top left), total consumer density (bottom left), final resource species richness (top right), final consumer species richness (bottom right), and initial consumer species richness in a predator–prey model. Consumers have additive effects on the per capita population growth rate of resources when $a = 0$, nonadditive antagonistic effects when $a < 0$, and nonadditive synergistic effects when $a > 0$. Modified from Ives et al. (2005).

diversity still contributes to enhance some ecosystem processes to some extent, but its effects are strongly dependent on the trophic level at which diversity varies and on the structure of the food web, in particular, its connectivity and the nature of population control (top-down vs. bottom-up). Nonlinear responses are common, except for special food-web structures, and negative effects at a very high diversity are possible because of collective overexploitation of resources by a diverse assemblage of efficient consumers. The prevalence of resource overexploitation in nature, however, is unknown and might be more limited than suggested by the models. Horizontal diversity does not prevent upper trophic levels from reducing the efficiency with which the ecosystem as a whole exploits the basal resource compared with a system with a single trophic level.

Despite the complexity introduced by trophic interactions, the relationships between biodiversity and ecosystem properties are predictable provided environmental conditions and food-web structure are known. Our model makes several predictions that deserve to be tested experimentally to gain better knowledge of the impacts of biodiversity changes on ecosystem functioning under natural conditions. Although a few experiments have been performed, they are still too limited to draw general conclusions regarding the functional effects of interactions between horizontal diversity and vertical diversity in food webs (Duffy et al. 2007).

Experimental changes in the diversity of a single trophic level within the context of a multitrophic system are slightly more frequent. Experimental manipulations of consumer diversity often showed enhanced resource exploitation and increased consumer biomass (Duffy et al. 2003, 2007), just as in single-trophic-level systems (chapter 3). There has been no evidence so far for the negative effects of consumer diversity that are predicted by the model under conditions conducive to resource overexploitation. This lack of evidence, however, might be due to the fact that recent experiments generally have not provided the necessary conditions for these effects to occur, in particular, sufficient time for prey abundance to adjust to changes in consumer diversity and feed back on consumer abundance. Antagonistic interactions among predators, which limit the potential for overexploitation, also appear to be relatively frequent (Schmitz 2007).

The converse effects of prey diversity on prey consumption by consumers have been studied much more frequently, although controlled experimental manipulation of prey diversity has been rare. Our model makes a straightforward prediction about these effects: since the presence of inedible plants allows part of the plant trophic level to escape consumption, herbivores are expected to have a smaller effect on plant biomass as plant diversity increases. Plants have a larger total biomass when some plant species experience no or reduced consumption (compare left and right panels in figure 4.8). Hillebrand and Cardinale (2004) performed a meta-analysis of a large number of experiments that manipulated the presence of invertebrate or vertebrate grazers while also measuring the magnitude of grazer effects on algal biomass and the diversity of algal assemblages. They found a consistent pattern of decreased consumer effects on algal biomass (i.e., consumer effects become less negative) as algal diversity increases (figure 4.11), in agreement with theoretical predictions.

Complete inedibility of some prey species, however, is not necessary for these effects to occur. Provided some mechanism other than top-down control by predators maintains prey diversity, differential susceptibility to predation coupled with differential dominance may often be sufficient to decrease predator effects on prey biomass. This is the basis for what is known as the dilution effect in the literature on disease ecology. The *dilution effect* occurs when increased species diversity reduces disease risk (Keesing et al. 2006). Striking examples of this effect include increased rice resistance to blast disease following increased rice genetic diversity (Zhu et al. 2000) and reduced risk of human exposure to Lyme disease as vertebrate host

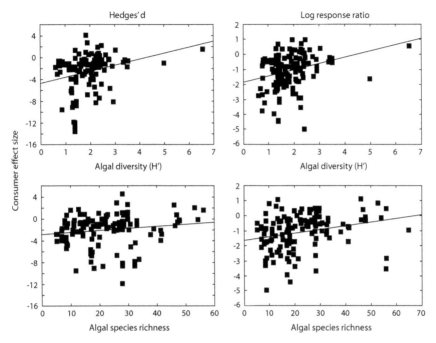

FIGURE 4.11. Consumer effect sizes on algal biomass vs. algal diversity in experiments that manipulated the presence of invertebrate or vertebrate grazers. Effects sizes are Hedges's d (left panels) or log response ratio (right panels). Algal diversity is measured by Shannon's diversity index (top panels) or species richness (bottom panels). Modified from Hillebrand and Cardinale (2004).

diversity increases (Ostfeld and Keesing 2000). The mechanism underlying the latter example is relatively well known and has been studied theoretically: as vertebrate host diversity decreases, the host species that transmit the disease most effectively (in this case, mice and deer) become dominant, thereby increasing disease risk. Although a model with random species loss would predict decreased Lyme disease risk on average as more species are lost, all realistic extinction scenarios predict that mice and deer will be the last species to go extinct, yielding an opposite trend toward increased Lyme disease risk as more species are lost (figure 4.12).

NONTROPHIC INTERACTIONS, BIODIVERSITY, AND ECOSYSTEM FUNCTIONING

While understanding the effects of trophic interactions on biodiversity and ecosystem functioning is challenging, an even greater challenge is to

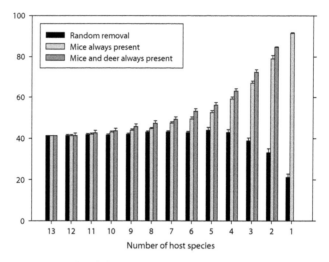

FIGURE 4.12. Results of simulations assessing the effects of reduced species richness on Lyme disease risk as measured by nymphal tick infection prevalence. Data are means ± 1 SE of 100 replicates. Three scenarios of species extinction are shown: (1) random species removal; (2) white-footed mice were present in all communities, but otherwise removal was random; and (3) mice and white-tailed deer were present in all communities, but otherwise removal was random. Reprinted from Ostfeld and LoGiudice (2003).

understand how *nontrophic interactions* affect biodiversity and ecosystem functioning. Trophic interactions and food webs have been studied abundantly from the early days of ecology, but nontrophic interactions have traditionally been neglected, especially by ecological theory (Bruno et al. 2003). The only exception is interference competition, since classical competition theory based on the Lotka–Volterra model does not differentiate between interference and exploitation competition. Recent mechanistic developments of competition theory, however, have mainly focused on exploitation competition, which is an indirect interaction resulting from the direct trophic interactions between consumers and their shared resources. I briefly discussed some specific effects of nontrophic interactions among consumers on the relationship between total resource or consumer biomass and consumer species richness in the previous section. Here I consider a wider range of nontrophic interactions that can potentially affect any species and any nontrophic interaction in ecosystems.

Recent experiments suggest that nontrophic interactions, such as facilitation, may play an important role in ecosystem functioning (Mulder et al. 2001; Cardinale et al. 2002; Rixen and Mulder 2005) and that different kinds of species interactions do not act in isolation but co-occur within the same community (Callaway and Walker 1997). Evidence for the importance of trait-mediated indirect interactions is also accumulating (Werner and Peacor 2003; Schmitz et al. 2004). Models of mutualism are often fairly specific and consider only one kind of species interaction. Simple models of mutualism also have the unrealistic property of leading to unlimited population growth under some conditions because they do not respect the principle of mass conservation. To explore interactions between community and ecosystem properties, ecological theory needs flexible, general ecosystem models that are able to include all types of direct species interactions (interference competition, mutualism, exploitation, commensalism, amensalism), as well as their indirect effects, while at the same time satisfying mass-balance constraints.

Following the pioneering work of Arditi et al. (2005), Alexandra Goudard and I recently developed an *interaction-web* model that meets this need (Goudard and Loreau 2008). Our model expands upon the food-web model presented in the previous section by adding nontrophic interactions in the form of nontrophic modifications of trophic interactions—that is, each species is allowed to modify the trophic interaction between any two species (figure 4.13). There are two main differences between this model and the food-web model described in the previous section. First, our interaction-web model includes a production efficiency, ε_x, for each species x. The amount of consumed nutrient that is not used in production is assumed to be recycled within the ecosystem. Second, and more fundamentally, the rate of consumption of species x by species y, c_{xy}, is now the product of a trophic component, the predation rate a_{xy}, and a nontrophic component, the nontrophic coefficient ν_{xy}:

$$c_{xy} = \nu_{xy} a_{xy}. \tag{4.15}$$

This nontrophic coefficient captures all the modifications of the trophic interaction between species x and y that are caused by the nontrophic effects of the $3S$ species (S species at each trophic level) in the ecosystem (including species x and y themselves). Each species z is allowed to modify the trophic interaction between species x and y; the size of this nontrophic effect depends on its biomass, X_z, and an interaction modification coefficient, μ_{xyz}:

$$\nu_{xy} = \prod_{z=1}^{3S} (1 + X_z)^{\mu_{xyz}}. \tag{4.16}$$

The function that describes nontrophic effects [equation (4.16)] was chosen such that it satisfies several conditions. First, it is a strictly increasing function of both the intensity of interaction modification, μ_{xyz}, and

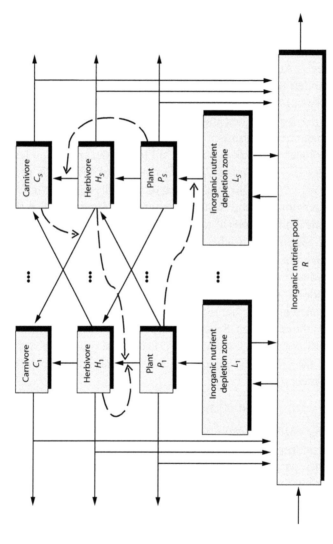

FIGURE 4.13. A complex nutrient-limited interaction web made up of S plants, S generalist herbivores, and S generalist carnivores, in which each species may modify trophic interactions between any two species through nontrophic effects. Plants have limited access to inorganic nutrient in individual resource depletion zones. Boxes represent nutrient stocks, thin solid arrows represent nutrient flows, and thick dashed arrows show examples of possible interaction modifications.

biomass, X_z. Second, if either $\mu_{xyz} = 0$ or $X_z = 0$, then $(1 + X_z)^{\mu_{xyz}} = 1$, and species z does not affect the trophic interaction between species x and y. In the absence of any interaction modification, $\nu_{xy} = 1$, and the consumption rate reduces to a simple trophic predation rate; i.e., $c_{xy} = a_{xy}$. Third, the function for nontrophic effects is strictly positive, so that the sign of the consumption rate does not change. The intensity of interaction modification, μ_{xyz}, can be either positive or negative while keeping the nontrophic coefficient, ν_{xy}, and hence the consumption rate, c_{xy}, positive. This ensures that, whatever the sign of the nontrophic effects of other species, the nutrient flow between species x and y is not reversed, and the food-web structure of the system remains intact. Last, interaction modification coefficients are symmetrical ($\mu_{xyz} = \mu_{yxz}$) to maintain mass balance.

In the presence of interaction modifications, the consumption rate c_{xy} can be smaller or larger than the corresponding trophic predation rate a_{xy} depending on whether the nontrophic coefficient ν_{xy} is smaller or larger than 1 [equation (4.15)], which in turn depends on whether the various species have negative or positive interaction modification coefficients μ_{xyz} [equation (4.16)]. Consequently, each species can either increase or decrease the population growth rate of any other species through nontrophic effects, so that all types of species interactions (competition, mutualism, exploitation, commensalism, amensalism) are incorporated in the model, including intraspecific density dependence (if μ_{xzz} or $\mu_{zyz} \neq 0$). Thus, our model describes a full interaction web. It also satisfies mass balance: interaction modifications change the material flow between a resource and a consumer, but what is gained by the consumer is lost by the resource, and vice versa, so that there is mass conservation overall.

We used this model to analyze the relationships between community and ecosystem properties that emerge from the assembly dynamics of complex ecosystems through successive species invasions from a regional species pool. Here I highlight a few of the main results that came out of this study. First, despite continuous species replacement due to invasion of new species, community and ecosystem properties stabilize relatively quickly in a *quasistationary regime*. In this regime, local species richness increases almost linearly with regional species richness (i.e., the number of species in the regional pool) despite the presence of strong species interactions. This confirms the prediction that species interactions generally do not limit local species diversity but only reduce it relative to the regional species pool (Loreau 2000a).

Second, a number of ecosystem properties, such as total biomass, plant biomass, carnivore biomass, plant production, herbivore production, carnivore production, and inorganic soil nutrient use, generally increase with regional species richness and hence also with local species richness. The positive effects of species diversity on these ecosystem properties, however, tend to level off at high levels of regional species richness in the presence of nontrophic interactions (figure 4.14). Surprisingly, biomass and production are typically lower in the presence of nontrophic interactions—i.e., in interaction webs—than in their absence—i.e., in simple food webs. Herbivore biomass is usually unaffected by species richness, which suggests a top-down control of carnivores on herbivores.

At first sight, one would expect facilitative and mutualistic interactions to be fostered by the presence of nontrophic effects, and these positive interactions to make the ecosystem more efficient. This is indeed what Arditi et al. (2005) found overall with their model. So, why do nontrophic interactions counterintuitively tend to reduce biomass and production at all trophic levels in our model? The answer is paradoxical: *they do so precisely because resource exploitation becomes more efficient.* The frequency and strength of nontrophic interactions can be easily manipulated in our interaction-web model by varying two parameters in the regional species pool: nontrophic connectance (the proportion of realized nontrophic effects among all possible nontrophic effects) and maximal nontrophic intensity (the maximum absolute value of the interaction modification coefficients). Increasing either of these parameters does increase the frequency of nontrophic species interactions, including mutualistic interactions, but concurrently it increases the mean resource exploitation ability of each species. This increased resource consumption leads to overexploitation, intense competition, and reduced resource-use complementarity at consumer trophic levels, which cascades down the food web and eventually results in decreased biomass and production at all trophic levels.

Two main differences between our model and that of Arditi et al. may explain why this outcome was not apparent in their study. First, they used relatively low levels of trophic connectance among species from different trophic levels, whereas we allowed all species to be generalist consumers. As we saw in the previous section, consumer generalism can greatly increase the potential for resource overexploitation. Second, nontrophic interaction modifications combined additively in their model, whereas they combine multiplicatively in our model [equation (4.16)]. As a result, nontrophic effects can increase resource consumption more strongly in our model, thereby further enhancing the potential for resource overexploitation. It is currently difficult to assess which of the two models is closer to reality for lack of appropriate empirical data. The two models highlight different potential outcomes that might occur in different ecosystems.

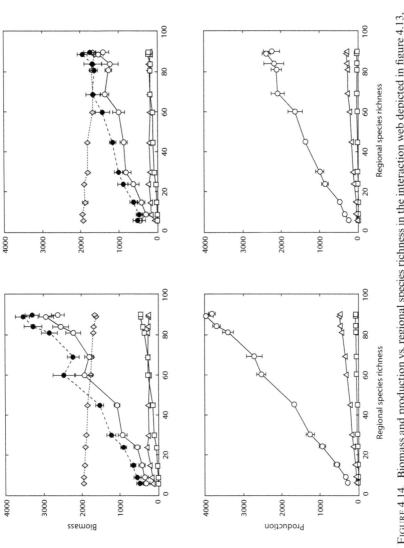

FIGURE 4.14. Biomass and production vs. regional species richness in the interaction web depicted in figure 4.13, in the absence (left panels) or presence (right panels) of nontrophic interactions. Results are means ±1 SD in the quasi-stationary regime for total biomass (•), total plant (○), total herbivore (△), and total carnivore (□) biomass and production. Dotted lines represent the amount of inorganic nutrient (◇). Modified from Goudard and Loreau (2008).

Thus, while positive species interactions such as facilitation and mutualism are one of the biological mechanisms that contribute to generate functional complementarity, and hence positive relationships between species diversity and total biomass and production within single trophic levels (chapter 3), their impact in multitrophic systems is more complex because they tend to increase the resource exploitation ability of species at all trophic levels. Consequently, they can enhance the efficiency with which limiting resources are used and transferred along the food chain, thereby contributing to enhance ecosystem functioning, but they can also exacerbate the negative effects of trophic interactions when consumers are generalists, including the potential for overexploitation, intense resource and apparent competition, and reduced functional complementarity at consumer trophic levels. Since a higher species diversity provides more opportunities for both trophic and nontrophic interactions, it can also exacerbate these negative effects and result in weaker, or even negative, relationships with total biomass or production at high diversity.

It is important to keep in mind, however, that our model did not explore all possible forms of nontrophic interactions or all possible scenarios for the topology and strength of these interactions. In particular, it focused on nontrophic modifications of resource consumption rates and assumed no restriction on either consumer generalism or maximal consumption rates. Other scenarios and other forms of nontrophic interactions that affect other demographic and functional parameters are likely to yield different results. But our model reveals the real potential for counterintuitive effects arising from nontrophic interactions, which are too often assumed a priori to be positive.

CONCLUSION

The general conclusion that emerges from this chapter is that trophic and nontrophic interactions make the relationships between biodiversity and ecosystem functioning more complex in ecosystems with multiple trophic levels than in the simple competitive systems that have been usually studied experimentally. Theory predicts that vertical diversity does not maximize ecosystem properties at the scale of trophic levels but instead makes them converge on intermediate values through damped oscillations as food-chain length increases. More integrative measures of ecosystem functioning, such as total ecosystem biomass, may even decrease as food-chain length increases. Overexploitation of biological resources is another factor that can cause deterioration of the functioning of diverse, strongly interacting eco-systems. Although horizontal species diversity is still expected to enhance ecosystem properties under a range of conditions, it can also reduce them when it promotes overexploitation. The fact that recent experiments have generally found positive effects of species diversity on total biomass and resource use at all trophic levels (chapter 3) does not invalidate these theoretical predictions because most of these experiments have manipulated the diversity of a single trophic level and have probably not created conditions conducive to resource overexploitation.

The models I have discussed, however, do not imply that the relationships between biodiversity and ecosystem functioning should always be complex in nature. Mutual interference, intraguild predation, and spatial heterogeneity may dampen top-down effects of consumers on their resources and thereby reduce the potential for overexploitation. Trade-offs between species traits are ubiquitous and can strongly affect the impacts of biodiversity on ecosystem processes. For instance, trade-offs between a consumer's ability to exploit a wide range of prey species and its ability to exploit any particular prey efficiently are likely. Our models so far have also assumed a constrained trophic structure with a limited number of distinct trophic levels (plant, herbivore, and carnivore), but omnivory and ontogenetic diet shifts are also common in nature. The effects of all these factors deserve more thorough investigation based on empirical data to predict the expected relationships between biodiversity and ecosystem properties in different types of ecosystems.

One of the most robust properties of food webs and interaction webs is that species additions or deletions can trigger abrupt changes in the structure and functioning of ecosystems. Such abrupt changes occur, for instance, when an entire trophic level is added or removed, or when an inedible species is added or removed. In both cases, the nature of the factors that control the various trophic levels may change, with major effects on the allocation of energy, materials, and biomass among them. This property makes trophic and nontrophic species interactions an important source of surprises and uncertainty in a rapidly changing world.

It seems fair to say that theory has still barely scratched the surface of complex interaction webs. One of the main strengths of the mechanistic approach followed in this chapter is that it allows simple principles to be revealed based on the operation of a limited set of elementary processes. But its corresponding limitation is that it cannot explore the full range of the possible. In particular, I have considered only systems that are limited by a single nutrient, an assumption that severely constrains species coexistence and ecosystem processes. It is my hope that the principles derived based on these constraints will help in understanding the properties of more complex ecosystems. But a more integrative approach that takes into account multiple limiting factors, multiple mechanisms of coexistence, and more realistic configurations of species interactions will be needed to fully account for the complexity and functioning of natural ecosystems. Merging the perspectives of food webs, nontrophic interaction webs, biodiversity, and ecosystem functioning remains an exciting challenge that is key to understanding and predicting future changes in natural and managed ecosystems and the services they provide to humans.

APPENDIX 4A
EFFECTS OF VERTICAL DIVERSITY IN A FOOD CHAIN

The effects of vertical diversity on the biomass, production, and ecological efficiency of the various trophic levels at equilibrium in the model food chain described by equations (4.1)–(4.4) can be studied by comparing the equilibrium values provided in table 4.1 and using the feasibility and invasion conditions for the various trophic levels. This analysis assumes that all equilibria are feasible and stable, which requires that environmental fertility I be high enough to support the top trophic level but not too high to avoid consumer satiation and destabilization of the system.

1. One Trophic Level ($N = 1$)
The persistence of the first trophic level ($B_{1(1)}^* > 0$) requires

$$B_{0(1)}^* < \frac{I}{m_0} = B_{0(0)}^*. \tag{4A.1}$$

2. Two Trophic Levels ($N = 2$)
Invasion by the second trophic level of a food chain with a single trophic level at equilibrium ($dB_2/dt > 0$ when $B_2 \approx 0$) requires

$$f_2(B_{1(1)}^*) - m_2 > 0,$$
$$B_{1(1)}^* > f_2^{-1}(m_2) = B_{1(2)}^*. \tag{4A.2}$$

g_2 (table 4.1) is a decreasing function of $B_{0(2)}^*$, hence its inverse, g_2^{-1}, which determines $B_{0(2)}^*$, is a decreasing function of $B_{1(2)}^*$. Therefore, because of inequality (4A.2),

$$B_{0(1)}^* < B_{0(2)}^*. \tag{4A.3}$$

The persistence of the second trophic level ($B^*_{2(2)} > 0$) also requires

$$B^*_{0(2)} < \frac{I}{m_0} = B^*_{0(0)}. \tag{4A.4}$$

Combining (4A.3) and (4A.4) yields

$$B^*_{0(1)} < B^*_{0(2)} < B^*_{0(0)}. \tag{4A.5}$$

Based on the expressions for the ecological efficiencies at equilibrium (table 4.1), this inequality implies that

$$\lambda^*_{1(2)} < \lambda^*_{1(1)}. \tag{4A.6}$$

Last, since Φ^*_1 is proportional to λ^*_1. [equation (4.5)], one also has

$$\Phi^*_{1(2)} < \Phi^*_{1(1)}. \tag{4A.7}$$

Although the biomass, production, and ecological efficiency of the first trophic level all decrease upon addition of the second trophic level [inequalities (4A.2), (4A.6), and (4A.7)], it is easy to show that its production and ecological efficiency decrease less than does its biomass. Indeed, $\Phi_1 = f_1(B_0)B_1$ [equation (4.3)]. Since B_0 and B_1 vary in opposite directions, the production, and hence also the ecological efficiency, of the first trophic level vary less than does its biomass.

3. THREE TROPHIC LEVELS ($N = 3$)

An analysis based on the same principles can be performed when there are three trophic levels, ultimately leading to the full inequalities (4.6)–(4.11).

The only indeterminacy concerns secondary (herbivore) production. From equation (4.5),

$$\Phi_2 = I\lambda_1\lambda_2. \tag{4A.8}$$

Since λ_1 and λ_2 vary in opposite directions upon addition of the third trophic level, Φ_2 can potentially vary in both directions. Note, however, that at equilibrium λ_1 and λ_2 are simple functions of the mass of inorganic nutrient, B_0, irrespective of the number of trophic levels. Thus, secondary production at equilibrium, Φ^*_2, is also a continuous function of B^*_0:

$$\Phi^*_2 = \varepsilon_1\varepsilon_2(I - m_0 B^*_0)\left[1 - \frac{m_1}{f_1(B^*_0)}\right] \tag{4A.9}$$

$$\frac{d\Phi^*_2}{dB^*_0} = \frac{\varepsilon_1\varepsilon_2}{f_1^2(B^*_0)}\{m_1 f_1{}'(B^*_0)(I - m_0 B^*_0) - m_0 f_1(B^*_0)[f_1(B^*_0) - m_1]\} \tag{4A.10}$$

In this equation, the term $I - m_0 B^*_0$ is positive and measures the net supply of inorganic nutrient available for consumption by the first trophic level (chapter 1), while the term $f_1(B_0^*) - m_1$ is also positive and measures the net productivity of the first trophic level available for consumption by the second trophic level. Thus, the variation of Φ_2^* as a function of B_0^* depends on the relative magnitude of the consumption flows by the first two trophic levels. It also depends on the form of the functional response of the first (plant) trophic level as determined by f_1 and its derivative f_1'. If the ecosystem reaches an equilibrium such that B_0^* is sufficiently low and lies in the steeply ascending part of the plant functional response, Φ_2^* will tend to vary in the same direction as does B_0^*; i.e., it will tend to decrease upon addition of the third trophic level. In contrast, if the ecosystem reaches an equilibrium such that B_0^* is sufficiently high and plant nutrient uptake is near saturation, $f_1'(B_0^*)$ will be close to zero, and Φ_2^* will vary in a direction opposite to B_0^*; i.e., it will increase upon addition of the third trophic level.

4. WHOLE-ECOSYSTEM PROPERTIES

The dynamics of the total nutrient stock, $B_T = \sum_{i=0}^n B_i$, in the ecosystem is obtained simply by summing equations (4.1) across all trophic levels, yielding

$$\frac{dB_T}{dt} = I - \sum_{i=1}^n \left(\frac{1}{\varepsilon_i} - 1\right) f_i(B_{i-1}) B_i - \sum_{i=0}^n m_i B_i. \qquad (4A.11)$$

Assume first that all trophic levels have maximal production efficiencies, i.e., $\varepsilon_i = 1$. The second term on the right-hand side of equation (4A.11) is then zero. Solving this equation at equilibrium yields

$$\sum_{i=0}^n m_i B_i^* = I. \qquad (4A.12)$$

The summation term on the left-hand side of this equation can be expressed in terms of the means and covariance of m_i and B_i^* across trophic levels. Therefore,

$$B_T^* = n.\overline{B^*} = \frac{I}{\overline{m}} - \frac{n.\text{cov}(m,B^*)}{\overline{m}}. \qquad (4A.13)$$

When the mass-specific loss rates of all trophic levels are equal ($m_i = m$), the total nutrient stock is simply $B_T^* = I/m$, a constant that is independent of food-chain length. But in principle it can be larger or smaller than this constant when the mass-specific loss rates of the various trophic levels differ, depending on the sign of the covariance between m and B^*.

Now relax the unrealistic assumption that all trophic levels have maximal production efficiencies and let $\varepsilon_i < 1$. The second term on the right-hand side of equation (4A.11) then becomes negative. Using the same derivation as above, it is straightforward to see that the total nutrient stock at

equilibrium is then necessarily smaller than that provided by equation
(4A.13) because of the additional negative term that has to be subtracted
from the right-hand side of this equation. As a result, the total nutrient
stock generally decreases as food-chain length increases (unless there is a
large negative covariance between m and B^*).

Since total ecosystem biomass at equilibrium is simply $B_T{}^* - B_0{}^*$ and
the inorganic nutrient stock $B_0{}^*$ alternates between smaller and larger values depending on the number of trophic levels in the system, total ecosystem biomass is expected to show the same overall decreasing trend with
food-chain length as does the total nutrient stock.

Total cumulative ecosystem production across all trophic levels at equi-
librium, $\Phi_T^* = \Sigma_{i=1}^n \Phi_i^*$, can be obtained as follows. Since model (4.1) tracks
nutrient stocks and flows and plants generally fully use the amount of
limiting nutrient they take up, $\varepsilon_1 \approx 1$. Hence primary production $\Phi_1{}^* =
I - m_0 B_0{}^*$, and the ecological efficiency of plants $\lambda_1{}^* = (I - m_0 B_0{}^*)/I$. Pro-
duction at higher trophic levels is then readily obtained using equation
(4.5). Summing production over all trophic levels yields

$$\Phi_T^* = (I - m_0 B_0^*)\left(1 + \lambda_2^* + \lambda_2^*\lambda_3^* + \cdots + \prod_{i=2}^n \lambda_i^*\right). \tag{4A.14}$$

When the ecological efficiencies of all consumer trophic levels are equal,
$\lambda_i{}^* = \lambda^*$, and food-chain length is large enough ($n \to \infty$), this equation simplifies to

$$\Phi_T^* = \frac{I - m_0 B_0^*}{1 - \lambda^*}. \tag{4A.15}$$

Since consumer ecological efficiencies are typically on the order of 2 percent to 10 percent, equations (4A.14) and (4A.15) show that total cumulative ecosystem production is expected to increase slightly overall as food-
chain length increases. Variations in the inorganic nutrient stock, $B_0{}^*$, due
to changes in top-down control, however, may override the production increments of the additional trophic levels.

APPENDIX 4B
EFFECTS OF NUTRIENT ENRICHMENT IN A FOOD CHAIN

The effects of nutrient enrichment on the biomass, production, and ecological efficiency of the various trophic levels at equilibrium in the model
food chain described by equations (4.1)–(4.4) can be studied by taking the
derivative of the equilibrium values provided in table 4.1 with respect to
environmental fertility I.

When the number of trophic levels $n = 0$ or 1, the results are straightforward and reported in table 4.2. When $n = 2$, the derivative of $B_{0(2)}^*$ with respect to I is obtained by implicit differentiation of the following equation derived from the dynamical equation for the inorganic nutrient at equilibrium:

$$I - m_0 B_{0(2)}^* - \frac{B_{1(2)}^* f_1(B_{0(2)}^*)}{\varepsilon_1} = 0. \tag{4B.1}$$

Since $B_{1(2)}^*$ is constant, implicit differentiation of this equation yields

$$\frac{dB_{0(2)}^*}{dI} = \frac{\varepsilon_1}{\varepsilon_1 m_0 + B_{1(2)}^* f_1'(B_{0(2)}^*)}, \tag{4B.2}$$

which is positive.

The sign of the variation of all the other variables follows immediately, except that of $\lambda_{1(2)}^*$, whose derivative is

$$\frac{d\lambda_{1(2)}^*}{dI} = \frac{\varepsilon_1 m_0 B_{1(2)}^* [B_{0(2)}^* f_1'(B_{0(2)}^*) - f_1(B_{0(2)}^*)]}{I^2 [B_{1(2)}^* f_1'(B_{0(2)}^*) + \varepsilon_1 m_0]}. \tag{4B.3}$$

The sign of $d\lambda_{1(2)}^*/dI$ depends on the form of the plant functional response, f_1. If the plant functional response is linear, as in a Lotka–Volterra interaction (type 1), $f_1' = f_1/B_0$ and $d\lambda_{1(2)}^*/dI = 0$. If the plant functional response is concave down (type 2, or the second part of type 3), $f_1' < f_1/B_0$ and $d\lambda_{1(2)}^*/dI < 0$. If it is concave up (the first part of type 3), $f_1' > f_1/B_0$ and $d\lambda_{1(2)}^*/dI > 0$. Since plant functional responses are usually of type 2, $\lambda_{1(2)}^*$ is expected to decrease more often than increase after nutrient enrichment.

When $n = 3$, there is no explicit solution for the equilibrium biomasses of trophic levels 0, 1, and 3. A logical argument, however, allows us to conclude that $\Phi_{1(3)}^*$, $B_{1(3)}^*$, and $B_{3(3)}^*$ necessarily increase with I. If I increases, either $B_{0(3)}^*$ or $B_{1(3)}^*$ (or both) must increase to compensate for this increased nutrient input, and hence primary production, $\Phi_{1(3)}^*$, which is an increasing function of $B_{0(3)}^*$ and $B_{1(3)}^*$, must increase. This increased inflow at the first trophic level must in turn be balanced by an increase in the sum of the outflows, i.e., plant mortality and secondary production. Since $B_{2(3)}^*$ is top-down-controlled and stays constant, this implies that $B_{1(3)}^*$ increases. As a result, $B_{3(3)}^*$, which is positively related to $B_{1(3)}^*$, also increases.

The sign of the variation of $B_{0(3)}^*$ can be determined by implicit differentiation of the following equation derived from the dynamical equation for the first trophic level at equilibrium:

$$B_{1(3)}^* f_1(B_{0(3)}^*) - m_1 B_{1(3)}^* - \frac{B_{2(3)}^* f_2(B_{1(3)}^*)}{\varepsilon_2} = 0. \tag{4B.4}$$

Implicit differentiation of this equation yields, after some algebra,

$$\frac{dB^*_{0(3)}}{dI} = \frac{dB^*_{1(3)}}{dI} \left\{ \frac{B^*_{2(3)}[B^*_{1(3)} f'_2(B^*_{1(3)}) - f_2(B^*_{1(3)})]}{\varepsilon_2 B^{*2}_{1(3)} f'_1(B^*_{0(3)})} \right\}. \tag{4B.5}$$

The sign of $dB^*_{0(3)}/dI$ depends on the form of the herbivore functional response, f_2. If the herbivore functional response is linear (type 1), $f'_2 = f_2/B_1$ and $dB^*_{0(3)}/dI = 0$. If it is concave down (type 2, or the second part of type 3), $dB^*_{0(3)}/dI$ has the opposite sign to $dB^*_{1(3)}/dI$ and hence is negative. If it is concave up (the first part of type 3), $dB^*_{0(3)}/dI$ has the same sign as $dB^*_{1(3)}/dI$ and hence is positive. Since herbivore functional responses are more likely to be concave down than concave up because of digestion limitations at high food availability, $B^*_{0(3)}$ is expected to decrease more often than increase after nutrient enrichment.

The sign of the variation of the other variables follows immediately from the above results.

APPENDIX 4C
ASSEMBLY RULE FOR TWO-LEVEL FOOD WEBS WITH SPECIALIST HERBIVORES

This appendix shows that the assembly rule depicted in figure 4.6 holds for the model food web described by equations (4.14).

In the case of a single food chain, this system is identical that studied in appendix 4A. Thus, for the food chain composed of plant 1 and herbivore 1, we have, from equation (4A.5),

$$B^*_{0(0,11)} < B^*_{0(0,11,21)} < B^*_{0(0)}. \tag{4C.1}$$

If plant 2 is added to this food chain without its specialist herbivore, it is limited only by the inorganic nutrient and hence eventually controls the inorganic nutrient stock at its own B^*_0 value:

$$B^*_{0(0,1,21,12)} = f^{-1}_{12}(m_{12}) = B^*_{0(0,12)}. \tag{4C.2}$$

On the other hand, plant 1 is top-down-controlled by herbivore 1 just as in the simple food chain:

$$B_{11(0,11,21,12)} = f^{-1}_{21}(m_{21}) = B^*_{11(0,11,21)}, \tag{4C.3}$$

while herbivore 1 is bottom-up-controlled by the nutrient left over by plant 2:

$$B^*_{21(0,11,21,12)} = \frac{[f_{11}(B^*_{0(0,12)}) - m_{11}] \varepsilon_{21} B^*_{11(0,11,21)}}{m_{21}}. \tag{4C.4}$$

The persistence of herbivore 1 requires

$$f_{11}(B^*_{0(0,12)}) - m_{11} > 0,$$

$$B^*_{0(0,12)} > f_{11}^{-1}(m_{11}) = B^*_{0(0,11)}. \tag{4C.5}$$

Whatever the food-web configuration, mass balance for the inorganic nutrient also imposes

$$m_0 B^*_0 + \sum_j f_{1j}(B^*_0)B^*_{1j}/\varepsilon_{1j} = I. \tag{4C.6}$$

The left-hand side of this equation is a monotonic increasing function of B^*_0. Since plant 2 introduces an additional term in this function while B^*_{11} is unchanged compared with the system without plant 2 [equation (4C.3)], one necessarily has

$$B^*_{0(0,11,21,12)} < B^*_{0(0,11,21)}. \tag{4C.7}$$

Combining (4C.1), (4C.5), and (4C.7),

$$B^*_{0(0,11)} < B^*_{0(0,11,21,12)} = B^*_{0(0,12)} < B^*_{0(0,11,21)} < B^*_{0(0)}. \tag{4C.8}$$

If we now add herbivore 2 to this system comprising plant 1, herbivore 1, and plant 2, both plants are top-down-controlled by their specialist herbivore at the same level as in a simple food chain, and both herbivores are bottom-up-controlled by the inorganic nutrient:

$$B^*_{2j(0,11,21,12,22)} = \frac{[f_{1j}(B^*_{0(0,11,21,12,22)}) - m_{1j}]\varepsilon_{2j}B^*_{1j(0,1j,2j)}}{m_{2j}}, \tag{4C.9}$$

which imposes

$$B^*_{0(0,11,21,12,22)} > f_{1j}^{-1}(m_{1j}) = B^*_{0(0,1j)} \tag{4C.10}$$

for both $j = 1$ and 2.

Using again the mass-balance constraint (4C.6) and the fact that B^*_{11} is unchanged compared with the system with food chain 1, one also has

$$B^*_{0(0,11,21,12,22)} < B^*_{0(0,11,21)}. \tag{4C.11}$$

Combining (4C.8), (4C.10), and (4C.11), we finally get the multiple inequality depicted in figure 4.6.

Generalizing this inequality to more than two food chains is straightforward since the constraints that arise from herbivore persistence and mass balance remain the same. Thus, adding a third food chain requires that the equilibrium values of the inorganic nutrient stock with and without the third herbivore be comprised between the corresponding values for the system with two food chains, just as the second food chain requires them to be comprised between the corresponding values for a single food chain.

CHAPTER 5

Stability and Complexity of Ecosystems:

New Perspectives on an Old Debate

Research into the potential consequences of changes in biodiversity on ecosystem functioning and on the delivery of ecosystem services has been prominent in fostering cross-fertilization between community ecology and ecosystem ecology during the last decade. This research has shown that biodiversity loss can have adverse effects on the average rates of ecosystem processes such as primary production and nutrient retention in temperate grassland ecosystems (chapter 3). Most of the evidence for this conclusion, however, comes from relatively short-term theoretical and experimental studies under controlled conditions, which do not address the long-term sustainability of ecosystems. The last chapter (chapter 4) extended this body of theory to more complex food webs and interaction webs but focused again on their functioning under equilibrium conditions.

It is of considerable interest to further understand how biodiversity loss will affect long-term temporal patterns in ecosystem functioning. Will ecosystem functional properties and services become more variable and less predictable as species diversity is reduced? Are species-rich ecosystems more capable of buffering environmental variability and maintaining ecosystem processes within acceptable bounds than species-poor ecosystems? These are fundamental questions that have considerable implications for our ability to understand, predict, and manage ecosystems in a changing world. In this chapter I synthesize recent theory that seeks to answer these questions.

As a matter of fact, these questions address in a new form a long-standing debate in ecology about the relationship between the complexity and stability of ecological systems. The study of this relationship has had a long and controversial history (May 1973; Pimm 1984, 1991; McCann 2000). It is therefore useful to understand the ins and outs of this debate before attempting to provide fresh answers to these questions. Accordingly, I first briefly summarize the central components of this debate to identify their main limitations and the questions they left unresolved. I then present and discuss new theoretical developments on the relationship between biodiversity and ecosystem stability, first in simple competitive systems, and then in more complex food webs. I show how these new approaches offer a potential resolution of the old debate by clearly identifying and linking stability properties at the population and ecosystem levels, how they provide new insights into the mechanisms that underlie ecosystem stability, and how they generate new questions for empirical and experimental studies at the interface between population ecology and ecosystem ecology.

A BRIEF HISTORY OF THE STABILITY–COMPLEXITY DEBATE

The traditional view that permeated ecology in its early days held that complex, diverse natural ecosystems are inherently more stable than simple, or artificially simplified, systems. This view was articulated theoretically by such great names in ecology as E. P. Odum (1953), MacArthur (1955), and Elton (1958), but it was so prevalent that it could be found in almost any ecology textbook in the 1950s and 1960s. Many arguments were used to support it, from philosophical standpoints about the perceived "balance of nature" contrasting with the disruptive influence of humans in natural systems, to theoretical or experimental evidence that simple model ecosystems are inherently unstable, through somewhat looser comparative empirical evidence that species-poor islands and artificial agricultural ecosystems are more prone to invasions by new species and pests than are their continental and natural counterparts. MacArthur (1955) also proposed, using a heuristic model, that the more pathways there are for energy to reach a consumer, the less severe will be the failure of any one pathway. This book is not the place to review and do justice to all these arguments, which have already been discussed by many others in the past. With hindsight, it is probably fair to say that many of these arguments were based on valid points, but that they were largely intuitive, remotely related to one another, and lacked a strong theoretical and experimental foundation. They became almost universally accepted because they represented the conventional wisdom described in the aphorism, "Don't put all your eggs in one basket."

Theoretical work by Levins (1970), Gardner and Ashby (1970), May (1972, 1973) and others challenged this traditional view in the early 1970s and eventually led to an almost diametrically opposite view regarding the stability of ecological systems. In particular, using a simple dynamical community model linearized in the vicinity of an equilibrium point and randomly drawn parameter values, May (1972) showed that the probability of having a locally stable equilibrium drops abruptly as the community crosses a threshold level of complexity. Specifically, local stability is almost certain when

$$\bar{\beta}\sqrt{SC} < 1, \tag{5.1}$$

while local instability is almost certain otherwise. In this inequality, S is the number of species in the system, C its connectance (proportion of nonzero species interactions among all possible interactions), and $\bar{\beta}$ the mean interaction strength (mean effect of a species' density on the per capita population growth rate of other species for all nonzero species interactions). May interpreted the left-hand side of this inequality as a measure of a system's complexity since it includes its diversity, connectance, and interaction strength. Inequality (5.1) then shows that complexity and diversity beget instability—not stability as previously believed.

In this work, stability was defined qualitatively by the fact that a system returns to equilibrium after a small perturbation. The intuitive explanation for the destabilizing influence of complexity is that the more diversified and the more connected a system, the more numerous and the longer the pathways along which a perturbation can propagate within the system, leading to either its collapse or its explosion. This conclusion was further supported by analyses of one quantitative measure of stability, resilience, in model food webs (Pimm and Lawton 1977; Pimm 1982).

This theoretical work was very influential, but it had a number of limitations. First, it was based on randomly constructed model communities. More realistic food-web models that incorporate thermodynamic constraints, partial donor control of trophic interactions, and observed patterns of interaction strengths do not necessarily have the same properties (DeAngelis 1975; de Ruiter et al. 1995; Brose et al. 2006; Neutel et al. 2007).

Second, stability is really a metaconcept that covers a wide range of different properties or components (table 5.1). Pimm (1984) recognized a number of these properties and concluded that the relationship between diversity and each of them need not be the same. Furthermore, each of these stability properties can be applied to a number of variables of interest at different hierarchical levels, such as individual species abundance, community species composition, and ecosystem properties (table 5.1). Again, the relationship between diversity and any stability property may be different for different variables (Pimm 1984; Ives and Carpenter 2007). This creates a large matrix of potential combinations of stability properties and

TABLE 5.1. Concepts and Definitions Related to Stability in Ecological Systems

Components of stability

Stability property	Definition
Qualitative stability	Property of a system that returns to its original state after a perturbation. Generally used for an equilibrium state, though it can also be applied to systems that return to nonequilibrium trajectories.
Resilience[a]	A measure of the speed at which a system returns to its original state after a perturbation (Webster et al. 1974). Generally used for an equilibrium state, though it can also be applied to systems that return to nonequilibrium trajectories.
Resistance	A measure of the ability of a system to maintain its original state in the face of an external disruptive force (Harrison 1979). Generally used for an equilibrium state.
Robustness	A measure of the amount of perturbation that a system can tolerate before switching to another state. Closely related to the concept of ecological resilience sensu Holling (1973). Can be applied to both equilibrium and nonequilibrium states.
Amplification envelope	Describes how an initial perturbation from an equilibrium state is amplified within a system (Neubert and Caswell 1997).
Variability	A measure of the magnitude of temporal changes in a system property. A phenomenological measure which does not make any assumption about the existence of an equilibrium or other asymptotic trajectories.
Persistence	A measure of the ability of a system to maintain itself through time. Generally used for nonequilibrium or unstable systems before extinction occurs.

Variables of interest

Individual species abundances
Species composition
Ecosystem properties

Sources of stability/instability

Internal: species interactions, demographic stochasticity
External: environmental changes, biological invasions, extirpations

[a] Some confusion surrounds the term "resilience" in the ecological literature. Though the term was first introduced into ecology by Holling (1973), it has most often been used in the sense defined by Webster et al. (1974). Here I follow the common usage without any judgment on the relative merits of the two definitions.

From Loreau et al. (2002a).

variables of interest, of which the new theory concerned only a small part. Specifically, May's (1972, 1973) and Pimm's (1982) theory essentially concerned the qualitative stability and resilience of communities as ensembles of populations, not the stability of ecosystem-level aggregate properties. Although May (1974) touched upon the difference between population- and community-level stability, he did not expand his exploratory work into a full-fledged theory.

Third, the formalism of autonomous, deterministic dynamical systems, which describes a fixed set of variables with time-independent parameters, inherently excludes a number of phenomena that characterize biological and ecological systems. In particular, it does not allow for the fact that these systems are subject to continuous environmental changes at various temporal scales and have the ability to react or adapt to these changes through asynchronous population dynamical responses, species replacement, phenotypic plasticity, and evolutionary changes. By ignoring these features, most of the theory on the complexity and stability of ecological systems has focused on deterministic equilibria and ignored much of the potential for functional compensation, both within and between species, which, as we shall see below, is the basis for the stabilization of aggregate ecosystem properties.

Despite these limitations, the view that diversity and complexity beget instability quickly became the new paradigm in the 1970s and 1980s because of the mathematical rigor of the theory. During this period, few dissenting voices were heard. Those proposing an alternative viewpoint were mostly ecosystem ecologists emphasizing functional compensation between species as the mechanism that stabilizes ecosystem processes against a background of wider variability of individual populations (Patten 1975; McNaughton 1977, 1993). Though often ignored, these ideas are the basis of the new wave of theoretical, experimental, and observational work that developed in the late 1990s, to which I now turn.

INSURANCE AND PORTFOLIO: BIODIVERSITY AS A STABILIZING FACTOR OF NATURE'S ECONOMY

Tilman (1996) provided the first experimental data suggesting that species diversity can simultaneously decrease population-level stability and increase community-level stability in grassland plant communities. These early results were controversial because confounding environmental factors drove variations in plant diversity in that experiment (Huston 1997).

But his team later obtained similar results in the Cedar Creek biodiversity experiment discussed in chapter 3, in which plant diversity was manipulated experimentally (figure 5.1). A number of empirical and experimental studies have reported similar stabilizing effects of species diversity on aggregate ecosystem properties.

Theory has since developed quickly, and in some areas outpaced experimental work, to account for such results that seemed to contradict theoretical efforts from the previous period. Two decisive features distinguish the new theoretical approaches to the diversity–stability relationship from earlier ones: first, these new approaches explicitly *differentiate and link stability properties at the population level and at the aggregate community or ecosystem level*; and second, they abandon the implicit assumptions that the environment is constant and that populations and ecosystems reach an equilibrium, to explicitly *incorporate population dynamical responses to environmental fluctuations*.

Following earlier conceptual contributions (Patten 1975; McNaughton 1977, 1993; Naeem 1998), three main approaches have been developed independently and simultaneously to analyze the effects of species diversity on the stability of community and ecosystem properties: (1) a statistical approach based on the phenomenological mean–variance scaling relationship, which considers neither population dynamics nor species interactions explicitly but which is easily applied to empirical data (Doak et al. 1998; Tilman et al. 1998; Tilman 1999); (2) a stochastic, dynamical approach that describes population dynamical responses to environmental fluctuations but does not explicitly consider species interactions (Yachi and Loreau 1999); and (3) a population dynamical approach that includes both a deterministic component describing species interactions and a stochastic component describing environmental fluctuations but that considers only small fluctuations in the vicinity of a deterministic equilibrium (Hughes and Roughgarden 1998, 2000; Ives et al. 1999; Ives and Hughes 2002). All three approaches have mainly focused on the temporal variability of aggregate community or ecosystem properties such as total biomass and total productivity in competitive communities. Temporal variability, as measured by the temporal variance or the temporal coefficient of variation, is a particularly useful stability property because it is easily measured empirically and it combines the effects of resistance and resilience (table 5.1).

The first two approaches generated two similar hypotheses known as the *portfolio effect* and the *insurance hypothesis*, respectively. Both hypotheses use metaphors borrowed from financial management and predict a stabilizing effect of biodiversity on aggregate ecosystem properties. The insurance

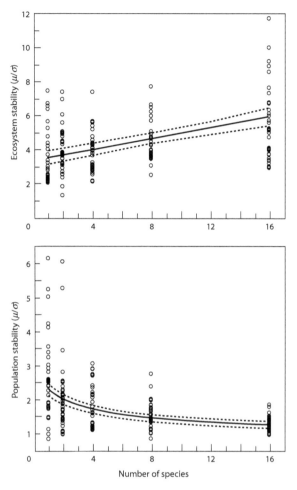

FIGURE 5.1. Community and population stability versus species richness in
the Cedar Creek biodiversity experiment. Top: community stability for the
decade from 1996 to 2005, measured by the ratio of mean plot total biomass
to its temporal standard deviation after detrending, is an increasing function
of the number of planted species. The regression line and its 95 percent confidence interval are shown. Bottom: plot-average population stability, determined with species biomass data for 2001–2005, is a declining function of the
number of planted species. The regression curve and 95 percent confidence
intervals are based on a fit of log(population stability) on log(number of species). Modified from Tilman et al. (2006).

metaphor emphasizes the role of biodiversity as a stabilizing factor of ecosystem functioning in the face of environmental fluctuations, while the portfolio metaphor emphasizes the behavior of the community that results from this stabilizing effect. The main contribution of the third approach has been to consider population dynamics and species interactions explicitly, and hence to provide more detailed insights into the way various factors interact to determine community stability.

The general mechanism that generates stabilization of aggregate ecosystem properties in diverse communities is simple in principle. Different species respond differently to their biotic and abiotic environment because of differences in their fundamental niche, thus generating *asynchrony in species environmental responses*. As their environment fluctuates through time, their abundance, biomass, and productivity also fluctuate through time in different ways that reflect these niche differences. Thus, differences in fundamental niches also yield differences in realized temporal niches, i.e., asynchrony of species fluctuations through time. Species asynchrony is the basis for *functional compensation* between species (McNaughton 1977): as one species decreases sharply in abundance, biomass, or productivity, another species decreases less sharply, or even increases, thus compensating partly or wholly for the decrease of the first species. As a consequence, the abundance, biomass, or productivity of the community as a whole fluctuates less than expected from individual species fluctuations (figure 5.2). The more species there are in the community and the more asynchronous their fluctuations, the larger the potential for stabilization of aggregate community or ecosystem properties (figure 5.2). Although the effects of species environmental responses, which capture fundamental niche differences between species, are mediated by the realized fluctuations in species abundance, biomass, or productivity, I shall show later that the key factor that allows understanding and predicting community stability is in fact the asynchrony of species environmental responses, not the asynchrony of realized population fluctuations.

The stochastic dynamical model that Shigeo Yachi and I developed (Yachi and Loreau 1999) revealed this mechanism clearly because it isolated it from the effects of other factors that drive population dynamics. Our model was based on two simple assumptions: (1) the productivity of each species obeys a stochastic process in response to environmental fluctuations, and (2) it fluctuates within the same range for all species (although this second assumption was chosen for convenience and can be relaxed easily). There was no restriction either on the probability density distributions of species responses or on within- and between-species temporal correlations

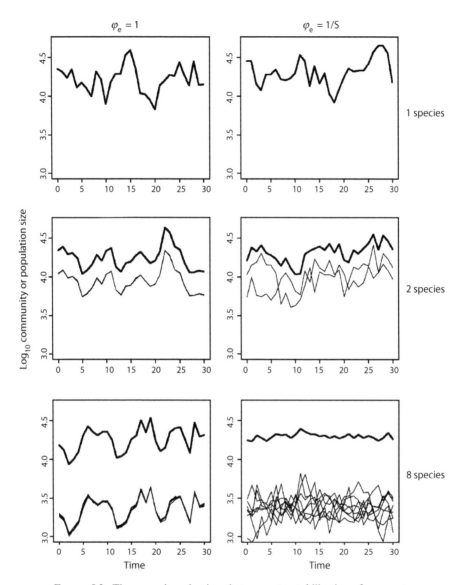

FIGURE 5.2. The general mechanism that generates stabilization of aggregate ecosystem properties in diverse communities. When species have asynchronous responses to environmental fluctuations (right panels, species have independent environmental responses, $\varphi_e = 1/S$), their population sizes (thin lines) also fluctuate asynchronously, which reduces the variability of community size (the sum of population sizes, thick lines). Increasing the number of species generally increases the potential for species asynchrony and hence the stabilization of community properties. When species have perfectly synchronous environmental responses (left panels, $\varphi_e = 1$), increasing the number of species does not contribute to stabilize community size. Simulated time series obtained using model (5.17) with $r_m = 0.5$, $K = 20{,}000$, $\alpha = 0$, and $\sigma_e = 0.3$.

in these responses. Total productivity at the ecosystem level at each time was then determined according to one of two rules: (1) determination by equivalence, in which interspecific competitive interactions are negligible, so that all species contribute equally to ecosystem productivity and the latter is simply the average of the various species' productivities; and (2) determination by dominance, in which interspecific competition is strong and ecosystem productivity is approximated by the productivity of the most productive species, as in the sampling effect (chapter 2). These two rules can be thought of as two limiting cases between which reality should generally lie.

The analysis of this model showed two types of biodiversity effects on ecosystem productivity in a fluctuating environment (figure 5.3): (1) a *stabilizing or buffering effect*, i.e., a reduction in the temporal variance (or other measures of variability) of ecosystem productivity; and (2) a *performance-enhancing effect*, i.e., an increase in the temporal mean of ecosystem productivity. Because species diversity contributes to maintain or enhance ecosystem functioning in the face of environmental fluctuations through both these effects, we called them *insurance effects* of biodiversity. The buffering effect generally occurs under both determination rules but disappears when there is perfect positive correlation between the various species responses (figure 5.3). Thus, its fundamental basis lies in the asynchrony of species responses, as explained above, rather than in the strength of competitive interactions. This asynchrony can be interpreted as a form of *temporal niche complementarity* between species. Since temporal niche differentiation promotes species coexistence (chapter 2), the conditions that promote coexistence within communities also promote the long-term stabilizing effects of biodiversity on aggregate community or ecosystem properties, just as they do for short-term biodiversity effects (chapters 2 and 3). The performance-enhancing effect is an additional effect that occurs under the rule of determination by dominance (figure 5.3). Its basis is that of the *selection effect*: biodiversity increases the range of trait variation available at any time, and a selective process such as interspecific competition promotes dominance by species that perform best under the current environmental conditions. This effect does not require complete dominance by the best-performing species; a slight selective advantage may suffice to generate it. Thus, the basic mechanisms involved in the insurance effects of biodiversity are very similar to those that operate in short-term biodiversity effects, i.e., temporal niche complementarity and selection of extreme trait values.

Our model showed that in principle any degree of species asynchrony, i.e., any deviation from perfect species synchrony, has the potential to stabilize aggregate community properties. Doak et al. (1998) and Ives et al.

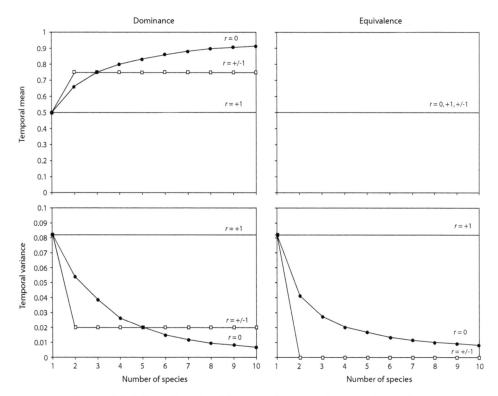

FIGURE 5.3. Effects of species richness on the expected temporal mean (top) and variance (bottom) of ecosystem productivity in a fluctuating environment in the two cases of determination by dominance (left) and determination by equivalence (right) in Yachi and Loreau's (1999) stochastic dynamical model. r is the correlation coefficient of species responses; $r = 0$, $+1$, and $+/-1$ correspond to the cases of independent responses, perfect positive correlation, and perfect negative correlation between two response functional groups ($r = +1$ within groups, $r = -1$ between groups), respectively. The probability density distribution is here assumed to be a uniform distribution on $[0, 1]$. Modified from Yachi and Loreau (1999).

(1999) reached the same conclusion with other approaches. This stabilization potential, however, may not be realized for at least two reasons. First, unequal species abundances or productivities tend to downplay functional compensation between species and hence stabilization of ecosystem properties (Doak et al. 1998; Yachi and Loreau 1999). In the extreme, if one species consistently dominates the community, other species make virtually no contribution to the magnitude and stability of aggregate community properties (Yachi and Loreau 1999). Second, although species diversity acts to decrease the variability of ecosystem properties through functional compensation between species, it may simultaneously increase the variability of individual populations because of the destabilizing effect of the increasing number of species interactions in the community. If the latter effect outweighs the former, their combination could in principle lead to a destabilization of ecosystem properties. Using population dynamical models of competitive communities, Ives and Hughes (2002) proved that, when interspecific competition is symmetrical (all interspecific competition coefficients are equal), destabilization of ecosystem properties occurs only if the variability of species per capita growth rates driven by environmental forcing increases with diversity. Since environmentally driven variability of species per capita growth rates is an individual-level property, there is no a priori reason to assume that it should change with the number of species. If this is the case, the insurance hypothesis should hold quite generally. Asymmetry of interspecific competition coefficients, however, tends to decrease the stabilizing effect of species diversity (Hughes and Roughgarden 1998); its effects are still to be studied quantitatively.

The statistical approach developed by Doak et al. (1998), Tilman et al. (1998), and Tilman (1999) has the advantage of being easily applied to empirical data, but it is more difficult to interpret because the mechanism underlying the stabilizing effect of diversity is hidden. To understand its properties and limitations, let us examine the basic model used by Tilman et al. (1998) and Tilman (1999). Denote the population size or biomass of species i at time t by $N_i(t)$, the total community size or biomass by $N_T(t) = \sum_{i=1}^{S} N_i(t)$, their respective temporal means by μ_{N_i} and μ_{N_T}, and their respective temporal variances by $\sigma_{N_i}^2$ and $\sigma_{N_T}^2$, which I shall henceforth call species variances and community variance, respectively.

Tilman et al.'s model makes a number of simplifying assumptions:

1. All species are assumed to obey the same constraints; in particular, they have the same temporal mean and variance of biomass.
2. The mean biomass of any species i, $\mu_{N_i} = m/S$, decreases in inverse proportion to the number of species, S, because of competition for shared resources, such that mean total biomass, $\mu_{N_T} = S\mu_{N_i} = m$, is independent of species richness.
3. The variance of species biomass scales with the mean such that

$$\sigma_{N_i}^2 = c\mu_{N_i}^z = cm^z S^{-z}. \qquad (5.2)$$

The scaling coefficient z equals 2 when variability, as measured by the standard deviation, is proportional to the mean, as is expected when processes are scale-independent (Doak et al. 1998). Empirical estimates of z in natural or experimental communities, however, typically lie between 1 and 2 (Tilman 1999; Steiner et al. 2005).

4. Species fluctuate independently of each other, such that the temporal covariance of their fluctuations in biomass is zero.

Tilman (1999) included an additional overyielding effect in his model to account for the positive effect of species richness on mean total biomass revealed by both theory and experiments (chapter 3), but this effect on the mean is of a different nature. Here I ignore it for the sake of simplicity and focus on the specific effects of diversity on the temporal variance. All the results and conclusions derived below, however, can easily be extended to the more complete model that includes the overyielding effect.

The variance of a sum of variables is the sum of the variances and covariances of all these variables. Therefore, community variance is

$$\sigma_{N_T}^2 = \Sigma \mathrm{var} + \Sigma \mathrm{cov}, \tag{5.3}$$

where $\Sigma \mathrm{var}$ is the summed species variances,

$$\Sigma \mathrm{var} = \sum_i \sigma_{N_i}^2, \tag{5.4}$$

and $\Sigma \mathrm{cov}$ is the summed species covariances,

$$\Sigma \mathrm{cov} = \sum_i \sum_{j \neq i} \mathrm{cov}(N_i, N_j). \tag{5.5}$$

Based on the above assumptions, community variance is simply

$$\sigma_{N_T}^2 = \Sigma \mathrm{var} = S\sigma_{N_i}^2 = cm^z S^{1-z}. \tag{5.6}$$

Thus, community variance declines as species richness increases provided $z > 1$, which is virtually always the case in empirical data. The same conclusion holds for the coefficient of variation—another commonly used measure of variability—since the coefficient of variation is the standard deviation divided by the mean and the mean is here assumed to be constant. Doak et al. (1998) used the term "statistical averaging" for this tendency for the variability of aggregated variables to decline for apparently purely statistical reasons, while Tilman et al. (1998) called it the "portfolio effect." Tilman (1999) and Lehman and Tilman (2000) contrasted this effect with the additional effect of negative summed covariances, which were supposed to encapsulate compensatory dynamics between species arising from interspecific competition. In this view, summed variances and summed covariances are interpreted as different mechanisms that contribute to the stabilizing effect of species diversity on community or ecosystem properties. The partition of community variance into these two statistical components has gained popularity because it can be applied easily to empirical data (e.g., Valone and Hoffman 2003; Gonzalez and Descamps-Julien 2004; Steiner et al. 2005; Tilman et al. 2006), and a growing number of authors have adopted the mechanistic interpretation of these components.

There are several fundamental problems, however, with this mechanistic interpretation of statistical relationships. The most fundamental problem is that statistical patterns in general are not enough to infer underlying causation or mechanisms. Concretely, in the present case, the statistical relationships expressed by equations (5.2), (5.6), and (5.7) do not have a clearly identified mechanism. As a consequence, negative summed covariances cannot be interpreted mechanistically either. Assumption (4) above is critical in this respect. Independence of species fluctuations may seem a reasonable intuitive null hypothesis representing a situation in which biological mechanisms such as interspecific competition and niche differentiation are absent, but it is not. Interspecific competition and niche differentiation are two counterbalancing processes, such that low niche differentiation generally implies relatively strong interspecific competition (relative to intraspecific competition), and high niche differentiation generally implies relatively weak interspecific competition (chapter 2). The only scenario in which the two processes vanish simultaneously is when competition (both within and between species) is absent altogether, i.e., in an ideal noninteractive community. This scenario, however, is precluded by assumption (2) above, which assumes strong competition that maintains total community biomass constant. Thus, the above model does not provide an internally consistent "null model" of a noninteractive community (Gotelli and Graves 1996), and none of its results can be interpreted as some sort of "statistical inevitability" (Doak et al. 1998). I shall show in the next section that the sign and magnitude of species covariances show complex relationships with the strength of interspecific competition and the amounts of temporal and nontemporal forms of niche differentiation. Therefore, their value cannot be assumed a priori.

It is straightforward to relax this unrealistic assumption of independent species fluctuations and extend the above statistical model to the general case where species covariances are not zero. Keeping the convenient assumption that all species have identical variances, summed covariances are then, by the definition of the correlation coefficient,

$$\Sigma\text{cov} = \sum_i \sum_{j \neq i} \rho_{N_i N_j} \sigma_{N_i}^2 = S(S-1) \overline{\rho_N} \sigma_{N_i}^2 = (S-1) \overline{\rho_N} \Sigma\text{var}, \qquad (5.7)$$

where $\rho_{N_iN_j}$ is the temporal correlation coefficient between the population sizes or biomasses of species i and j, and $\overline{\rho_N}$ is the average correlation coefficient between any two species in the community. Substituting this expression into equation (5.3) yields

$$\sigma_{N_T}^2 = cm^z S^{2-z} \varphi_N. \qquad (5.8)$$

In this equation, φ_N is a standardized communitywide measure of the synchrony of species abundances or biomasses, which is defined as

$$\varphi_N = \frac{\sigma_{N_T}^2}{\left(\sum_i \sigma_{N_i}\right)^2}. \qquad (5.9)$$

In this expression, the numerator is the observed community variance, while the denominator is the maximum value it could achieve were all species to fluctuate in perfect synchrony (Loreau and de Mazancourt 2008). This statistic has the advantage of being standardized between 0 (perfect asynchrony) and 1 (perfect synchrony) irrespective of the number of species. In contrast, the average correlation coefficient has a lower bound, $\overline{\rho}_{N_{min}} = -1/(S-1)$, that increases steadily with the number of species, thereby making comparisons among communities difficult to interpret (Loreau and de Mazancourt 2008).

In the special case where all species variances are equal, the dependence of this statistic on species richness, S, and on the average temporal correlation coefficient between species, $\overline{\rho}_N$, can be made explicit:

$$\varphi_N = \frac{1 + (S-1)\overline{\rho}_N}{S}. \qquad (5.10)$$

Thus, communitywide synchrony increases with the average temporal correlation between species. It stays constant at its minimum value of 0 when the average correlation is at its minimum value, $\overline{\rho}_{N_{min}} = -1/(S-1)$, and stays constant at its maximum value of 1 when the average correlation is also maximum ($\overline{\rho}_N = 1$). But it decreases with species richness for any intermediate value of the average correlation when the latter is kept constant. In particular, it declines as $1/S$ in the special case where species fluctuate independently ($\overline{\rho}_N = 0$).

Equation (5.8) shows that when variability is scale-independent ($z = 2$), any deviation from perfect synchrony ($\varphi_N < 1$) is sufficient to ensure a stabilizing effect of species richness on community biomass, in agreement with the conclusions of other approaches (Ives et al. 1999; Yachi and Loreau 1999). But when variability is scale-dependent, as it often is in empirical data, species richness stabilizes community biomass only when the scaling coefficient z is sufficiently large and the average correlation coefficient is sufficiently low (appendix 5A). When the scaling coefficient $z \leq 1$, species richness always destabilizes community biomass.

A hidden and poorly appreciated feature of the statistical model is that its various parameters are not independent of each other. To see this, consider a community obeying "neutral" community dynamics (Hubbell 2001), i.e., a community of equivalent species in which strong competition occurs but is identical within and between species. If all species are equivalent, their number and identity should have no effect on aggregate community properties, in particular, total biomass and its variance. Community variance is then constant, and equal to the variance of the biomass of a single species; i.e.,

$$\sigma_{N_T}^2 = cm^z. \tag{5.11}$$

Substituting this expression into equation (5.8) yields the following constraint that binds the scaling coefficient and species synchrony together:

$$\varphi_N = S^{z-2}, \tag{5.12}$$

or, equivalently,

$$z = 2 + \frac{\ln \varphi_N}{\ln S}. \tag{5.13}$$

These equations show that, for a given species richness, a specific value of the scaling coefficient necessarily implies a specific value of species synchrony, and vice versa. Furthermore, these values generally vary with species richness. At one extreme, if species fluctuate in perfect synchrony ($\varphi_N = 1$), variability should be scale-independent ($z = 2$). At the other extreme, if species fluctuate independently ($\varphi_N = 1/S$), variability should be proportional to the mean as in a Poisson process ($z = 1$). I shall show in the next section that these two limiting cases can arise from the action of different forces in population dynamics. Endogenous density dependence and exogenous environmental forcing are two forces that synchronize the population dynamics of equivalent species and hence contribute to generate the first limiting case ($\varphi_N = 1$, $z = 2$). By contrast, demographic stochasticity tends to make population fluctuations independent and hence contributes to generate the second limiting case ($\varphi_N = 1/S$, $z = 1$). Therefore, the combination of these forces is expected to generate intermediate values of both species synchrony and the scaling coefficient.

Deviations from these predictions occur when species are not equivalent, which provides an interesting set of predictions regarding the conditions under which species richness either stabilizes or destabilizes aggregate community properties (figure 5.4). Under the hypothesis of a neutral community of equivalent species, species richness has no effect on community variance. This occurs when species synchrony and the scaling coefficient constrain each other according to equations (5.12) and (5.13) (diagonal panels with gray background in figure 5.4). When species synchrony is smaller than expected in a neutral community, species richness stabilizes total community biomass (upper right panels with white background in figure 5.4). In contrast, when species synchrony is higher than expected in a neutral community, species richness destabilizes total community biomass (lower left panels with white background in figure 5.4). Note that summed variances and summed covariances bear no simple relation to community stability. For instance, summed covariances can be either positive or negative and can either increase or decrease with species richness in cases where species richness stabilizes total community biomass (upper right panels with white background in figure 5.4). These observations, as well as the hidden relationships between parameter values revealed above, show that mechanistic interpretations of the statistical approach are unwarranted.

The statistical and mechanistic approaches to the relationship between diversity and stability can be reconciled by noting that the mechanism of species asynchrony that underlies the insurance hypothesis also implicitly underlies the statistical averaging or portfolio effect. The portfolio effect emerges as the outcome of asynchronous species fluctuations in species-rich communities. In the same way, statistical averaging occurs in a well-managed portfolio because the latter contains a diversity of financial assets that fluctuate asynchronously. If assets are similar and subject to the same market forces and fluctuations, increasing the number of assets does little to reduce the fluctuations of the portfolio. Therefore, the insurance and portfolio hypotheses may be regarded as roughly equivalent.

Much confusion on this issue arose from an inconsistent application of the concept of statistical averaging. Statistical averaging is the statistical outcome of large numbers of individual events that occur at smaller scales or lower hierarchical levels and that tend to average out at larger scales or higher hierarchical levels (Patten 1975; McNaughton 1977, 1993). When the scales considered differ greatly (such as between particle physics and thermodynamics), small-scale events appear as essentially independent, random events at the larger scale because the laws that describe processes are different at the two scales and small-scale variations tend to cancel each other out at the larger scale. This is basically the idea that Doak et al. (1998) applied to the effect of species richness on variability of total community biomass. The main problem with this application, however, is that it takes independent species fluctuations for granted and does not explain where

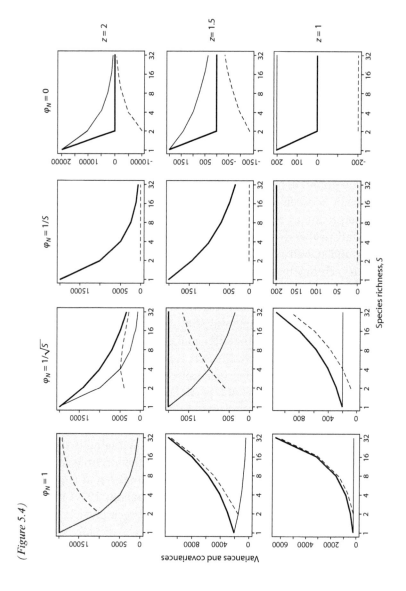

(*Figure 5.4*)

these come from in the first place. Unless one is dealing with a completely noninteractive community, species fluctuations will be approximately independent only if the number of species is sufficiently large and their environmental responses are asynchronous, as I shall show in the next section. Thus, the mechanism that stabilizes community properties, i.e., asynchronous environmental responses, is actually hidden in the assumption of independence of species fluctuations.

Statistical averaging is a useful property that can be used to predict some aggregate properties of large-scale systems, but it should be applied and understood critically. It cannot be used to predict how ecosystem properties vary with species richness as it was in the recent diversity–stability literature, since it applies to systems that have large numbers of microscopic components (species) in the first place. It cannot be used either to understand mechanisms at the microscopic scale as it was in the recent diversity–stability literature, since it attempts precisely to ignore these microscopic mechanisms. And its predictions break down when the conditions under which it is approximately valid are not met, i.e., when the number of species is small, when population fluctuations are large because of either exogenous environmental forcing or endogenous destabilizing density dependence, or when species show positively correlated responses to the environment—a range of conditions that probably include many natural communities. Thus, the concept of statistical averaging is much more restrictive than that of asynchronous species environmental responses because it assumes independent species fluctuations, an assumption that is only approximately

---

FIGURE 5.4. Community variance (thick solid lines), summed species variances (thin solid lines), and summed species covariances (thin dotted lines) versus species richness, S, for four values of species synchrony, φ_N, and three values of the scaling coefficient, z, in the statistical model (5.8). Diagonal panels with a gray background correspond to neutral communities with equivalent species that obey equations (5.12) and (5.13) and show no stabilization or destabilization of total biomass (community variance is constant). Upper right panels with a white background correspond to nonneutral communities in which species synchrony is smaller than expected in a neutral community, leading to a stabilizing effect of species richness on total biomass (community variance decreases). Lower left panels with a white background correspond to nonneutral communities in which species synchrony is higher than expected in a neutral community, leading to a destabilizing effect of species richness on total biomass (community variance increases). Summed species variances are confounded with community variance when species fluctuations are independent ($\varphi_N = 1/S$). Other parameter values: $c = 2$ and $m = 100$. From Loreau and de Mazancourt (unpublished results).

valid under restrictive conditions and that is unlikely to be met in most real communities. By contrast, asynchrony of species environmental responses is a general mechanism that applies to all kinds of communities.

SPECIES SYNCHRONY AND ECOSYSTEM STABILITY: A MECHANISTIC APPROACH

Given the intrinsic limitations of the statistical approach, a mechanistic approach is necessary to understand the factors that drive species asynchrony and their effects on the stability of community or ecosystem properties in multispecies communities. The population dynamical approach developed by Hughes and Ives (Hughes and Roughgarden 1998, 2000; Ives et al. 1999; Ives and Hughes 2002) provides a useful basis to develop such a mechanistic approach. Their analysis, however, was mainly focused on the dynamics of competitive communities driven by small environmental fluctuations in the vicinity of a deterministic equilibrium, and their models ignored demographic stochasticity. Claire de Mazancourt and I recently extended this approach to include demographic stochasticity and consider population fluctuations far from equilibrium (Loreau and de Mazancourt 2008, and unpublished results).

Our starting point is the development of a *neutral model* that describes the dynamics of a community in fluctuating environments in the absence of any form of niche differentiation. This first step is important because the null hypothesis against which community dynamics should be compared is unclear based on recent studies. The hypothesis of independent species fluctuations is often used, implicitly or explicitly, as a null hypothesis to test for the effects of biological mechanisms such as niche differentiation and interspecific competition on community dynamics (Frost et al. 1995; Doak et al. 1998; Tilman et al. 1998; Tilman 1999; Klug et al. 2000; Lehman and Tilman 2000; Ernest and Brown 2001; Houlahan et al. 2007). But it does not have any solid mechanistic basis, as I mentioned in the previous section. Hubbell's (2001) neutral model provides elegant predictions for species abundance patterns and fluctuations in saturated, space-limited communities, but it considers only population fluctuations driven by demographic stochasticity and ignores fluctuations driven by endogenous density dependence and exogenous environmental forcing, which are ubiquitous in natural communities.

Recent neutral models differ conceptually from traditional "null models" in ecology (Gotelli and Graves 1996). Traditional null models were designed as formalized null hypotheses against which the effects of inter-specific competition on community structure can be tested. Accordingly, these models seek to remove the effects of interspecific competition. In contrast, recent neutral models assume strong interspecific competition for space. Their distinctive feature is that they assume no role for niche differentiation among species; accordingly, they assume that all individuals are competitively equivalent.

Our neutral model extends this concept by including the three main forces that drive population dynamics, i.e., intra- and interspecific density dependence, environmental forcing, and demographic stochasticity. Assume a set of S equivalent species that are limited by a common limiting factor and that respond identically to environmental fluctuations. Let $N_i(t)$ be the population size of species i at time t, and $r_i(t) = \ln N_i(t + 1) - \ln N_i(t)$ be its instantaneous per capita population growth rate at time t. Further assume that community size, $N_T(t) = \sum_{i=1}^{S} N_i(t)$, is regulated according to a simple discrete-time logistic equation with intrinsic rate of natural increase r_m and carrying capacity K. The theory of stochastic population dynamics predicts that, to a first-order approximation, each species will then obey a dynamics described by the equation

$$r_i(t) = \ln N_i(t + 1) - \ln N_i(t) = r_m \left[1 - \frac{N_T(t)}{K} \right] + \sigma_e U_e(t) + \frac{\sigma_d U_{di}(t)}{\sqrt{N_i(t)}},$$

(5.14)

where σ_e^2 and σ_d^2 are the environmental and demographic variances, respectively, and $U_e(t)$ and $U_{di}(t)$ are independent normal variables with zero mean and unit variance (Lande et al. 2003; Engen et al. 2005).

Although community size is regulated, individual population sizes drift as a result of demographic stochasticity. For species that do not go extinct, however, it is possible to obtain the expected temporal variances and covariances of their per capita population growth rates, which are, respectively,

$$\sigma_{r_i}^2 = \sigma_c^2 + \sigma_e^2 + \sigma_d^2 / \tilde{N}_i,$$

(5.15)

$$\text{cov}(r_i, r_j) = \sigma_c^2 + \sigma_e^2,$$

(5.16)

where $\tilde{N}_i$ is the harmonic temporal mean of species i's population size, and $\sigma_c^2 = (r_m^2 / K^2) \sigma_{N_T}^2$ is the community response variance, defined as the temporal variance of per capita population growth rates due to regulation of community size (Loreau and de Mazancourt 2008).

There are three additive components to the temporal variances and covariances of the per capita population growth rates of equivalent species: (1) one component due to endogenous regulation of community size, σ_c^2;

(2) another component due to exogenous environmental forcing, σ_e^2; and (3) a third component due to demographic stochasticity, $\sigma_d^2/\tilde{N}_i$. When demographic stochasticity is weak compared with community regulation and environmental forcing ($\sigma_c^2 + \sigma_e^2 \gg \sigma_d^2/\tilde{N}_i$), species are expected to fluctuate synchronously [$\text{cov}(r_i, r_j) \approx \sigma_{r_i}^2$, yielding a correlation coefficient close to 1]. In particular, when the intrinsic rate of natural increase $r_m > 2$, endogenous density dependence yields cyclic or chaotic deterministic attractors and hence considerable fluctuations in community size that strongly synchronize population fluctuations. In contrast, when community regulation and environmental forcing are weak compared with demographic stochasticity ($\sigma_c^2 + \sigma_e^2 \ll \sigma_d^2/\tilde{N}_i$), species are expected to fluctuate independently [$\text{cov}(r_i, r_j) \ll \sigma_{r_i}^2$, yielding a correlation coefficient close to 0].

Equation (5.16) predicts that the covariances of the per capita population growth rates of all species pairs are positive and identical. This prediction differs radically from Hubbell's (2001) neutral model, which predicts negative covariances between species abundances. There are two reasons why the two models make such contrasting predictions. First, Hubbell's makes the stringent assumption that community size is constant and species abundances obey a zero-sum game. This assumption leads automatically to negative covariances between species abundances because variations in species abundances must compensate exactly for each other to yield a constant sum (figure 5.5A). In our neutral model, community size is allowed to vary under the influence of endogenous density dependence and exogenous environmental forcing. These changes affect all species simultaneously and hence tend to synchronize their population dynamics (figure 5.5B). The assumption of constant community size is probably appropriate for tree communities in which there is strong competition for space and recruitment is high enough to quickly fill gaps. But the assumption of variable community size is probably appropriate for a wide range of other communities in which competition for space is not so constraining. Our model provides alternative predictions for such communities.

Second, an important distinction needs to be made between *per capita population growth rates* and *population sizes*. When community size is kept constant, the two demographic variables yield similar negative correlations (figure 5.5, left). But when community size varies through time, per capita population growth rates are more strongly correlated than population sizes (figure 5.5, right) because their fluctuations capture the short-term effects of the forces that govern population dynamics from one generation to the next, including the synchronizing effects of community regulation and environmental forcing. By contrast, long-term fluctuations in population

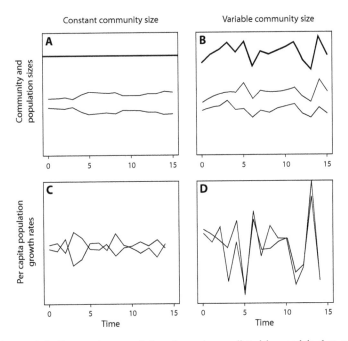

FIGURE 5.5. Contrasting population dynamics predicted by models that assume constant community size (left) and variable community size (right).
When community size (thick line) is kept constant, variations in species
abundances (thin lines) must compensate exactly for each other (correlation
coefficient = −1, A). Per capita population growth rates are then also strongly
negatively correlated (correlation coefficient = −0.99, C). Variations in community size, however, tend to synchronize species abundances (correlation
coefficient = 0.50, B), and even more so per capita populations growth rates
(correlation coefficient = 0.87, D). Left and right panels were obtained from
the same time series generated by model (5.17) where $r_m = 1$, $K = 1,000$,
$\alpha = 0.95$, $\sigma_e = 0.08$, and $\varphi_e = 0.94$, but absolute abundances were converted
into relative abundances to yield constant community size in left panels.

sizes are affected to a larger extent by ecological drift, which tends to desynchronize population fluctuations and plays a prominent role when species are equivalent. As a result, the synchrony of population sizes is affected by the length of the time series considered and is less predictable
than the synchrony of per capita population growth rates.

Our neutral model can easily be generalized to incorporate *niche differences* between species by relaxing the hypothesis of species equivalence in
two different ways: (1) by letting interspecific competition be smaller than
intraspecific competition, which generates a *nontemporal form of niche differentiation* that decouples density dependence in the various species; and

(2) by allowing species to have different responses to environmental forcing, which generates *temporal niche differentiation*. These two factors constitute deterministic sources of asynchrony that add to the effects of demographic stochasticity. The nonneutral version of model (5.14) reads

$$r_i(t) = r_m \left[1 - \frac{(1 - \alpha)N_i(t) + \alpha N_T(t)}{K'} \right] + \varepsilon_i(t) + \frac{\sigma_d U_{di}(t)}{\sqrt{N_i(t)}}. \tag{5.17}$$

This model is an extension of model (3.1) examined in chapter 3 in which the effects of environmental and demographic stochasticity on per capita population growth rates are added. As before, I assume for the sake of simplicity that all species have equal intrinsic rates of natural increase r_m, carrying capacities K', and interspecific competition coefficients α ($0 \le \alpha \le 1$). I also remove the effect of community size on variability by standardizing species carrying capacities such that the carrying capacity of the whole community, K, is independent of α:

$$K' = \frac{1 + \alpha(S - 1)}{S} K. \tag{5.18}$$

These simplifying assumptions allow exploring the specific role of niche differences between species as compared with the neutral baseline scenario, and removing the confounding effects of differences between species in competitive ability and variations in community size. The assumption of constant community size, however, can easily be relaxed. If species carrying capacities are kept constant whatever the strength of interspecific competition, community size varies accordingly, and most of the results derived below for community variance apply to the coefficient of variation of community size instead. This is easily understood as the coefficient of variation is but an indirect, a posteriori way to remove the effect of variations in the mean on the variance.

Environmental stochasticity is incorporated through $\varepsilon_i(t)$, which describes the environmental response of species i at time t. The environmental responses of the various species can now be more or less asynchronous. I assume for simplicity that the environmental variance, $\mathrm{var}(\varepsilon_i) = \sigma_e^2$, is identical for all species as before. A convenient measure of the synchrony of environmental responses, φ_e, is provided by the statistic of community-wide synchrony presented in the previous section applied to species environmental responses, i.e.,

$$\varphi_e = \frac{\mathrm{var}\left(\sum_i \varepsilon_i\right)}{S^2 \sigma_e^2} = \frac{1 + (S - 1)\overline{\rho_e}}{S}, \tag{5.19}$$

where $\overline{\rho_e}$ is the average correlation between species environmental responses.

This relatively simple but general model can be used to analyze not only the relationships between species diversity and community stability but also the mechanisms that underlie them and the way they are affected by the various factors that drive population dynamics. We performed these analyses in two complementary ways. First, we derived first-order analytical approximations for community variance and for the variances, covariances, and synchrony of both per capita population growth rates and population sizes. These approximations have limitations since they require sufficiently small population fluctuations, in particular, values of the intrinsic rate of natural increase that lead to a stable equilibrium of community size ($0 < r_m < 2$). But they provide useful analytical expressions that allow disentangling the effects of the various parameters on population and community stability. Second, we performed extensive numerical simulations of the model to analyze its properties when population fluctuations are larger, in particular, when the intrinsic rate of natural increase is larger and generates cyclic or chaotic asymptotic dynamics ($r_m > 2$). In our numerical simulations, however, we included demographic stochasticity in the form of a Poisson process, which is more realistic than the normal approximation used in equations (5.14) and (5.17) (Loreau and de Mazancourt 2008).

The first important question that can be examined using this model is how *niche differentiation affects species synchrony*. Intuitively, one may expect that temporal niche differentiation, in the form of a reduced synchrony of environmental responses, generates increasingly asynchronous species fluctuations. But, based on the views that prevail in the current literature, one may also expect that nontemporal niche differentiation, in the form of a reduced competition coefficient, has the opposite effect of synchronizing population fluctuations. A large number of recent studies have assumed, implicitly or explicitly, that interspecific competition should desynchronize the population fluctuations of competing species. This assumption has been the rationale for using negative summed covariances or related statistics as measures of compensatory dynamics in ecological communities (Frost et al. 1995; Tilman et al. 1998; Tilman 1999; Klug et al. 2000; Lehman and Tilman 2000; Ernest and Brown 2001; Houlahan et al. 2007).

First-order approximations provide straightforward predictions about the effects of temporal niche differentiation on species synchrony. Although the approximations of the synchrony of per capita population growth rates, φ_r, and of the synchrony of population sizes, φ_N, are fairly complex, in the limiting case when $\alpha = 0$ they both reduce to the simple expression (Loreau and de Mazancourt 2008)

$$\varphi_r|_{\alpha=0} \approx \varphi_N|_{\alpha=0} \approx \frac{\varphi_e \sigma_e^2 + \sigma_d^2/K}{\sigma_e^2 + \sigma_d^2 S/K}. \tag{5.20}$$

This equation predicts that species synchrony should vary between φ_e when environmental forcing is strong compared with demographic stochasticity ($\sigma_e^2 \gg \sigma_d^2/K$), and $1/S$ when environmental forcing is weak compared with demographic stochasticity ($\sigma_e^2 \ll \sigma_d^2/K$). Provided environmental forcing is not negligible, species synchrony should increase linearly with the synchrony of environmental responses, φ_e, when interspecific competition is absent. Numerical simulations confirm that species synchrony does generally increase with the synchrony of environmental responses, as expected intuitively, although the patterns are much more linear and smoother in the case of per capita population growth rates than in the case of population sizes.

Contrary to the prevailing view, *interspecific competition* does not desynchronize, but instead *synchronizes fluctuations in per capita population growth rates* (figure 5.6A and D). Although this result contradicts widely held beliefs, it makes sense intuitively for stronger interspecific competition means stronger coupling of density dependence in the various species. The effect of interspecific competition on the synchrony of population sizes, however, is more complex. When the intrinsic rate of natural increase is small, both the first-order approximation and numerical simulations predict that the synchrony of population sizes increases with the strength of interspecific competition when the synchrony of environmental responses is low (figure 5.6B) but decreases with the strength of interspecific competition when the synchrony of environmental responses is high (figure 5.6E). When the intrinsic rate of natural increase is large, the synchrony of population sizes shows a hump-shaped relationship with the interspecific competition coefficient (figure 5.6B and E). This pattern is the result of two counteracting factors. At first, increasing the strength of interspecific competition synchronizes population sizes, just as it does for per capita population growth rates, because it couples strong density dependence between species. But as the interspecific competition coefficient approaches 1, species become increasingly equivalent. Ecological drift then plays a major role, desynchronizing long-term fluctuations in population sizes despite the increased synchrony of short-term fluctuations in per capita population growth rates.

The second important question that can be examined using this model is how *niche differentiation affects community stability* and its relationship

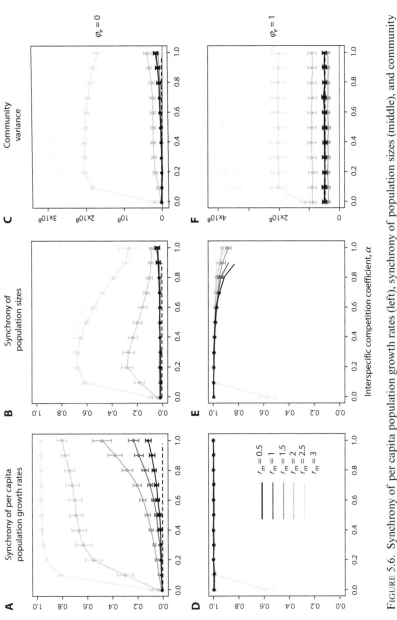

FIGURE 5.6. Synchrony of per capita population growth rates (left), synchrony of population sizes (middle), and community variance (right) (means ± 1 SD) versus the interspecific competition coefficient, α, for two values of the synchrony of environmental responses, φ_e, in the mechanistic model (5.17). The six curves in each panel correspond to different values of r_m, as indicated in panel D. Dashed lines (sometimes confounded with solid lines or with each other) show the corresponding first-order approximations for the three values of the intrinsic rate of natural increase that yield a stable equilibrium. Other parameter values: $S = 8$, $K = 20,000$, and $\sigma_e = 0.3$. From Loreau and de Mazancourt (2008, and unpublished results).

with species diversity. Recent studies have assumed that compensatory dynamics driven by species asynchrony constitutes the mechanism that generates community stability (Klug et al. 2000; Ernest and Brown 2001). Although this idea makes intuitive sense, the relationship between community stability and species asynchrony need not be so simple. As a matter of fact, community variance, and hence community stability, is a much simpler function of the factors that drive population dynamics than is species synchrony. Community variance is, to a first-order approximation (Loreau and de Mazancourt 2008),

$$\sigma_{N_T}^2 \approx \frac{K^2(\varphi_e \sigma_e^2 + \sigma_d^2/K)}{r_m(2 - r_m)}. \tag{5.21}$$

This equation predicts, as might be expected, that community variance increases linearly with the synchrony of environmental responses, just like the synchrony of per capita population growth rates and the synchrony of population sizes, at least when interspecific competition is absent. But it also predicts, counterintuitively, that *community variance is independent of the strength of interspecific competition, α*, when changes in community size are controlled for. This confirms Ives et al.'s (1999) conclusion based on a similar model without demographic stochasticity. The independence of community variance from the strength of interspecific competition stands in sharp contrast to the strong dependence of species synchrony on this parameter.

These predictions hold to a good approximation when population fluctuations are sufficiently small, i.e., when both the environmental variance and the intrinsic rate of natural increase are small (figures 5.6C and F and 5.7). When the environmental variance is large or when the intrinsic rate of natural increase exceeds the critical value that leads to limit cycles or chaotic dynamics ($r_m > 2$), however, community variance increases with the strength of interspecific competition, especially when the synchrony of environmental responses is low (figure 5.6C and F).

Another counterintuitive prediction of equation (5.21) is that *species richness affects community variance only indirectly, through the synchrony of environmental responses, φ_e*. Numerical simulations confirm this conclusion when population fluctuations are small: community variance is independent of species richness when the synchrony of environmental responses is kept constant (figure 5.7). Thus, once variations in community size are controlled for, species diversity can stabilize community properties only by decreasing the communitywide synchrony of environmental responses. Desynchronization of environmental responses occurs necessarily in multispecies communities because φ_e declines from its maximum value of 1

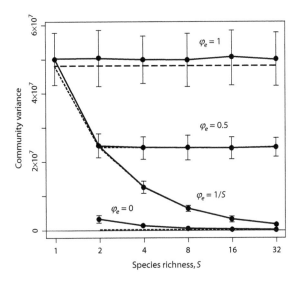

FIGURE 5.7. Community variance (mean $\pm$ 1 SD) versus species richness in the mechanistic model (5.17). The four curves correspond to different values of the synchrony of environmental responses, φ_e. By definition, $\varphi_e = 1$ when there is a single species; as a result, φ_e switches abruptly to a lower value when a second species is added, and φ_e is set to a fixed value <1. Dashed lines (sometimes confounded with solid lines) show the corresponding first-order approximations. Other parameter values: $r_m = 0.5$, $K = 20,000$, $\alpha = 0$, and $\sigma_e = 0.3$. From Loreau and de Mazancourt (unpublished results).

when a single species is present to $\overline{\rho}_e$ as the number of species increases [equation (5.19)]. This decline can be gradual when the average correlation between species environmental responses, $\overline{\rho}_e$, is zero or positive (as is the case when environmental responses are independent: figure 5.6), but it can be abrupt if environmental responses are strongly asynchronous. Maximal asynchrony of environmental responses occurs when $\overline{\rho}_e = -1/(S - 1)$. In this case, φ_e drops to its minimum value of 0 as soon as there are two species in the community and does not change anymore as the number of species further increases (figure 5.7). Note that these patterns are identical to those revealed by our earlier, much simpler stochastic model (figure 5.3).

The third and last question that can be examined using this model is whether the partition of community variance into summed species variances and summed species covariances is a useful tool to identify the mechanisms that underlie the stabilizing effect of species diversity on community properties. I have already alluded to some of the fundamental problems involved in the mechanistic interpretation of these statistical components in the previous section. Our model can be used to further examine their properties quantitatively.

In contrast to community variance, which has the remarkable property of being independent of the number of species and of the strength of interspecific competition when population fluctuations are sufficiently small, summed variances and summed covariances prove to be strongly affected by these parameters. They are also strongly interdependent since their sum, i.e., community variance, is independent of these parameters. Our mathematical and numerical analyses confirmed the results obtained in the previous section (figure 5.4); that is, both the sign and magnitude of summed covariances are extremely variable and difficult to interpret. Thus, *summed variances and summed covariances do not reveal the mechanisms at work in the stabilization of community properties.* In fact, they even obscure these mechanisms since they are strongly affected by factors that do not affect, or affect only weakly, community stability. Separating summed variances and summed covariances gives the misleading impression that the former are the result of individual species properties while the latter reflect species interactions. But summed variances are affected by species interactions just as much as are summed covariances. Their common dependence on species interactions is precisely what allows them to compensate for each other and yield a community variance that is almost independent from, or at least less dependent on, these interactions.

The main mechanism that drives the stabilizing effect of species diversity on aggregate community or ecosystem properties is the asynchrony of species environmental responses. This appears clearly from the first-order approximation of community variance [equation (5.21)]. The only factor that is affected by species richness in equation (5.21) is the synchrony of environmental responses, which declines as the number of species increases provided all species do not have perfectly correlated environmental responses. The strength of demographic stochasticity is unaffected by species richness when community size stays constant because it operates at the individual scale and hence depends only on the number of individuals in the community. Note, however, that the strength of demographic stochasticity is inversely proportional to community size. Therefore, when species diversity increases total biomass, as is generally observed in experiments (chapter 3), reduction in the strength of demographic stochasticity is another potential mechanism that can generate a stabilizing effect of species on aggregate ecosystem properties. In this case, the variance of community size increases, but its coefficient of variation [the square of which is obtained by dividing equation (5.12) by K^2] decreases.

An important distinction needs to be made between the *synchrony of species abundances*, φ_N, and the *synchrony of environmental responses*, φ_e. A species' environmental response describes the immediate response of its per capita population growth rate to exogenous environmental fluctuations. It expresses the species' fundamental niche. By contrast, a species' abundance is the result of its past abundance and of all the forces that affect its dynamics. The longer the temporal window considered, the wider the fluctuations in abundance because of ecological drift. Thus, fluctuations in abundance are only distantly related to a species' fundamental niche. A confusing factor here is that species environmental responses and their synchrony play an important part in driving species abundances and their synchrony, leading to partly similar variations. In particular, when the synchrony of environmental responses is zero, the synchrony of species abundances is also minimum [equation (5.20)]. In spite of these similarities, however, the two forms of synchrony obey partly different constraints and need to be carefully distinguished.

This conclusion suggests that recent studies of *compensatory dynamics* in natural communities (Frost et al. 1995; Ernest and Brown 2001; Klug et al. 2000; Houlahan et al. 2007) may need to be refocused. Historically, interest in compensatory dynamics arose from its putative role in ecosystem stability. But *functional compensation* is a term that probably better describes the ability of different species to stabilize aggregate functional processes through differential responses to environmental fluctuations. McNaughton (1977) was perhaps the first author to clearly argue for the role of functional compensation in community stability, and many of the examples he discusses concern differences in species functional responses to environmental changes, not so much changes in species abundances. By contrast, studies of compensatory dynamics generally focus on fluctuations in species abundances. These fluctuations are easy to measure, but unfortunately, as I have argued above, they are only distantly related to functional compensation and community stability.

Another, more technical problem with a number of recent studies is their reliance on negative summed covariances as a measure of compensatory dynamics. The intuitive concept of compensation implies that losses in one form are balanced by gains in another form. Perfect compensation is easily defined as a complete balance of losses and gains, in which case summed covariances are indeed necessarily negative (figure 5.5, left). This is easily seen by rewriting equation (5.3) in the form

$$\Sigma\text{cov} = \sigma^2_{N_T} - \Sigma\text{var}. \tag{5.22}$$

If compensation is perfect, community variance is zero and summed co-variances are equal to minus summed variances. But perfect compensation is an ideal case that is never achieved in nature. When community size varies through time, compensation can but be partial (figure 5.5, right). Summed covariances are then the difference between two positive terms in equation (5.22), and hence they can be either positive or negative. The benchmark for negative summed covariances is zero covariances, i.e., independent population fluctuations. But this benchmark is somewhat arbitrary since true independence is unlikely to occur in nature, as I have showed above. Therefore, negative summed covariances do not appear to have any strong logical or biological basis as a measure of partial compensatory dynamics. Given the limitations inherent in recent applications of the concept of compensatory dynamics, future research would benefit from focusing on functional compensation as a mechanism underlying ecosystem stability rather than on mere patterns of population fluctuations.

Two cautionary notes are worth making to close this section on community stability in competitive communities. First, one of the pieces of the stability jigsaw that is still missing here is the interconnection between community stability and the maintenance of species diversity due to temporal environmental variability. We saw in chapter 2 that some species can coexist precisely because of temporal variability, and we have now seen in this chapter that species that do coexist can contribute to stabilize ecosystem properties in temporally variable environments. One limitation of all the approaches considered here, however, is that species coexistence is either assumed a priori or built in through nontemporal forms of niche differentiation (yielding an interspecific competition coefficient smaller than 1). Examining the interactions between species diversity and community stability in communities where species coexistence is maintained specifically by temporal variability remains a future challenge.

Second, the theory presented here deals with temporal variability around a constant mean, that is, with stationary distributions. Tackling the effects of directional environmental changes, such as those of the current climate change, requires other approaches. Norberg et al. (2001) have started to explore how phenotypic trait diversity may affect the ability of ecosystems to respond to such directional changes on evolutionary time scales. Using an approach borrowed from quantitative genetics, they predict that phenotypic variance within functional groups is linearly related to the ability of these groups to respond to environmental changes. These results suggest that biodiversity can also serve as insurance against long-term directional environmental changes. On the other hand, biodiversity may also inhibit the evolutionary responses of individual species to changing environments because of the presence of competitors (de Mazancourt et al. 2008). Therefore, more work is needed on this important topic before final conclusions can be drawn.

BIODIVERSITY AS INSURANCE IN FOOD WEBS

The theory we have discussed so far concerns competitive communities with a single trophic level. Just as food-web structure affects the magnitude of ecosystem processes (chapter 4), we should expect it to also affect their variability and stability. As a matter of fact, classical theories and empirical studies on the relationships between diversity and stability have mostly concerned food webs or interaction webs, in which a richer array of species interactions are likely to propagate perturbations through ecosystems, thereby offering a greater potential for destabilization.

Despite this additional complexity, Ives et al. (2000) showed that, under conditions that keep food-web structure constant (the community is a combination of identical modular "subcommunities"), as well as the combined strength of species interactions on any given species constant independently of species diversity, the main conclusions obtained for competitive systems should also hold for multitrophic systems; i.e., (1) the strength of the interactions between different subcommunities should not affect the variability of aggregate community properties, and (2) increasing species richness should increase the stability of aggregate community properties provided different species respond differently to environmental fluctuations. Combined interaction strength is defined here as the combined per capita effects of all species on the population growth rate of a given species (as measured by the coefficients of the Jacobian matrix). This result was obtained for small random fluctuations in the vicinity of an equilibrium.

Elisa Thébault and I extended this analysis by relaxing the assumption of constant combined interaction strength and examining how different food-web configurations affect the relationships between species diversity, interaction strength, and the stability of ecosystem properties in ecosystems with two trophic levels, plants and herbivores (Thébault and Loreau 2005). We did this by studying two different models: (1) a discrete-time predator–prey model similar to that of Ives et al. (2000), which allows analytical treatment; and (2) the nutrient-limited ecosystem model previously used to study the relationship between diversity and the magnitude of ecosystem processes in multitrophic systems (figure 4.7), which allows more realism using numerical simulations. In both models, we compared two cases of variation of competition intensity with diversity, depending on whether the net strength of competition (1) increases or (2) stays constant with diversity. These two cases were crossed with the same three food-web configurations as in chapter 4; i.e., (1) herbivores are strictly specialists; (2) herbivores are generalists that compensate for the loss of plant species by increasing their consumption rate on other species (i.e., there is a trade-off between herbivore generalization and predation rate such that the voracity of each herbivore is independent of diversity); and (3) herbivores are generalists, and their consumption rate on each plant species is independent of plant diversity (i.e., there is no trade-off between herbivore generalization and predation rate such that herbivore voracity increases with diversity). We also analyzed the effects of species diversity on both species-level variability (coefficient of variation of biomass of individual plant and herbivore species) and aggregate community-level variability (coefficient of variation of total biomass at each trophic level).

Here I present only a few typical results obtained for the nutrient-limited ecosystem model in which environmental fluctuations are described by sinusoidal fluctuations of temperature and plant and herbivore mortality rates are Gaussian functions of temperature with different degrees of niche differentiation (figure 5.8). The main conclusions that emerge from these results are the following. First, species diversity acts generally to reduce the variability of total biomass at the various trophic levels, thus increasing ecosystem-level stability, irrespective of food-web structure (figure 5.9A and B). One significant exception to this pattern, however, occurs when herbivores are generalists and there is no trade-off between herbivore generalization and predation rate (figure 5.9C). By contrast, species diversity usually decreases population-level stability, although increased population-level stability is also possible (figure 5.9D–F). Thus, our model shows that *the insurance hypothesis also applies to multitrophic systems* and that increased ecosystem-level stability is often accompanied by decreased population-level stability.

Second, *temporal niche differentiation* also has opposite effects on ecosystem-level stability and population-level stability: while it tends to increase ecosystem-level stability (figures 5.9A and B), it generally decreases population-level stability (figure 5.9D–F). Temporal niche differentiation decreases population-level stability because direct and indirect competitive interactions between species tend to amplify the fluctuations of populations that are out of phase and slow down their return to equilibrium, just as in some competitive systems (Abrams 1976).

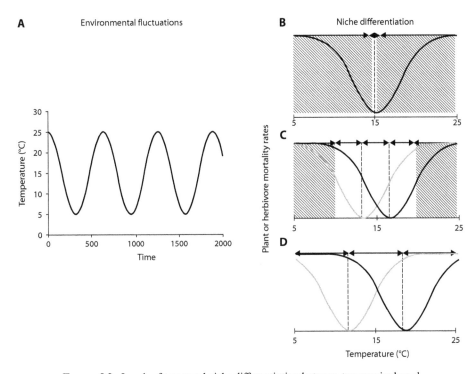

FIGURE 5.8. Levels of temporal niche differentiation between two species based on fluctuations of temperature incorporated in the nutrient-limited multitrophic ecosystem model depicted in figure 4.7. (A) Environmental fluctuations are sinusoidal with a temperature amplitude of 10°C, a mean of 15°C, and a period of 200π. (B) First degree of differentiation: niches are all centered on the average temperature of the system. (C) Second degree of niche differentiation: minimum values of mortality rates are regularly distributed over half the temperature gradient. (D) Third degree of differentiation: minimum values are distributed over the whole temperature gradient. Hatched areas correspond to the parts of the temperature gradient that are not accessible to minimum values. Modified from Thébault and Loreau (2005).

Third, *food-web connectivity per se* has few effects on ecosystem-level stability (compare figure 5.9A and B). By contrast, it has a strong stabilizing effect on population fluctuations when species have asynchronous environmental responses (compare figure 5.9D and E). This stabilizing effect of food-web connectivity at the population level is consistent with MacArthur's (1955) prediction that generalist predators should be buffered against asynchronous variations in their resources. Figure 5.9 makes clear, however, that this prediction does not extend to the ecosystem level.

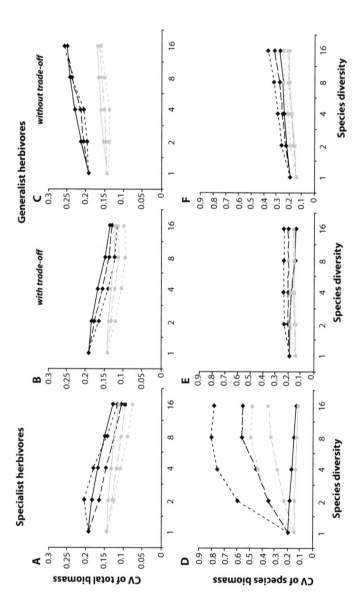

FIGURE 5.9. Coefficients of variation (CVs) of total and species-specific plant and herbivore biomass (± 1 SE) versus species diversity in the nutrient-limited multitrophic ecosystem model depicted in figures 4.7 and 5.7 for three food-web configurations. In each panel, black lines with diamonds correspond to CVs for plants, while gray lines with squares correspond to CVs for herbivores. Solid lines correspond to the first degree of niche differentiation of plants and herbivores, dashed lines correspond to the second, and dotted lines correspond to the third. Modified from Thébault and Loreau (2005).

Last, the key factor that determines the stabilizing or destabilizing effect of species diversity on ecosystem-level stability is *consumers' combined interaction strength*. When herbivores are generalists and there is no trade-off between herbivore generalization and predation rate, combined interaction strength increases with diversity. In this case, both population-level stability and ecosystem-level stability decrease as species diversity increases (figure 5.9C and F). This was implicitly the case in earlier theoretical studies by May (1972, 1973), Pimm (1982), and many others since interaction coefficients were assumed to be constant irrespective of the number of species or trophic links. But consumers' combined interaction strength may often stay constant, or even decrease, with prey species diversity, in which case our model shows that diversity has a stabilizing effect on ecosystem properties. One mechanism that can generate a trade-off between herbivore generalization and predation rate is that prey diversity forces predators to spend more time on information processing, thereby reducing prey consumption and decreasing trophic interaction strength (Kratina et al. 2007). The dilution effect in disease transmission (chapter 4) can have similar effects.

By clearly distinguishing two hierarchical levels at which stability properties can be different and explicitly analyzing the key role played by trade-offs and interaction strength in the stability–diversity relationships, these new theoretical developments have the potential to reconcile the seemingly contradictory results of previous approaches and provide more precise predictions based on the biology of interacting organisms. In particular, it is noteworthy that both MacArthur's (1955) and May's (1972, 1973) results, which have been traditionally opposed, are obtained with the same model and sometimes even under the same conditions. This emphasizes once more that the diversity–stability relationship is a complex, multifaceted one that does not lend itself to sweeping statements. The most critical insight provided by recent models, however, is the potential for a stabilizing influence of species diversity on ecosystem properties in food webs just as in simple competitive systems, provided either consumers are specialized or there is a cost to their being generalists.

Few experimental studies have manipulated species diversity in food webs and examined stability at the two hierarchical levels. Steiner et al. (2005) reported results for aquatic food webs that are consistent with the new theory. In a laboratory-based multitrophic aquatic microcosm containing bacteria, algae, heterotrophic protozoa, and rotifers, the variability of total biomass declined as mean realized species diversity increased, as expected when net interaction strength does not increase with diversity (figure 5.10). The variability of population-level biomass, however, also

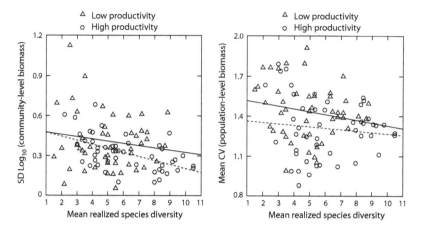

FIGURE 5.10. Temporal variability of community-level (left panel) and population-level (right panel) biomass vs. mean realized species richness in a laboratory-based multitrophic aquatic microcosm. The solid line shows the linear regression for low-productivity treatments; the dashed line shows the linear regression for high-productivity treatments. Modified from Steiner et al. (2005).

decreased slightly with diversity, an outcome that is not expected generally but that is predicted by our model under some circumstances. Kolasa and colleagues (Kolasa and Li 2003; Romanuk and Kolasa 2004; Vogt et al. 2006) also found positive effects of species diversity on population-level stability in multitrophic rock pool systems after controlling for some confounding factors. These studies suggest that, paradoxically, we might now have a stronger theory for ecosystem-level stability than for population-level stability. The generality and mechanistic basis of these results, however, still need to be assessed.

Although significant, these recent advances linking population- and ecosystem-level variability do not exhaust the topic of the stability of complex food webs. A number of stability properties other than variability are still poorly understood (Ives and Carpenter 2007), and the influence of a number of structural and dynamical features of food webs on their stability is only beginning to be uncovered. McCann and colleagues (McCann et al. 1998; McCann 2000; Rooney et al. 2006) have recently emphasized the importance of *weak interactions* and *structural asymmetry* in the stability of food webs. The stabilizing effect of weak interactions is explicit in May's equilibrium theory [equation (5.1)]. But McCann et al. (1998) focused on a different stabilizing effect of weak interactions; i.e., weak interactions have the potential to increase the persistence of simple food webs by dampening the destabilizing effect of strong consumer–resource interactions that generate cyclic or chaotic dynamics in the system. Assume that a generalist predator consumes two prey species, a preferred prey with which it interacts strongly and another prey with which it interacts weakly. If the first predator–prey interaction is strong enough to generate sustained oscillations, the two prey species will tend to covary negatively because the second prey will flourish whenever the first is suppressed by high consumer densities. In turn, the first, preferred prey will be released from strong consumption pressures when the second prey increases in density, thus preventing it from reaching very low densities. Structural asymmetry between fast and slow energy channels from different prey species to a shared top predator has the same effect of dampening internally generated predator–prey oscillations (Rooney et al. 2006).

This elegant theoretical work proposes a new avenue to tackle food-web stability. But it has not yet been used to study the stability of aggregate ecosystem properties, and it is still unclear to what extent the mechanism on which it hinges explains the stability of real food webs. First, natural food webs have many more species and interactions than the simplified strong-weak, fast-slow channels pictured in this hypothesis. Second, many natural food webs do not show clear evidence of cyclic or chaotic dynamics induced by strong predator–prey interactions. Third, this hypothesis assumes that population fluctuations are entirely driven by these strong species interactions and ignores the ubiquitous influence of environmental forcing on natural communities. Vasseur and Fox (2007) have recently shown that environmentally induced fluctuations in the mortality rates of intermediate consumers contribute to synchronize their dynamics, as expected from the theory developed in the previous section, and that this synchronization of consumer dynamics can paradoxically promote food-web stability because of the transient responses of basal resources and top predators. Thus, more work is needed to assess the contribution of this mechanism to the stability of natural ecosystems. A recent microcosm experiment, however, suggests that this mechanism might explain the positive effect of species diversity on population-level stability in ecosystems with multiple trophic levels (Jiang et al. 2009).

One robust lesson that emerges clearly from this work, from the recent work on biodiversity and ecosystem stability reviewed above, and from classical work on food-web stability in the vicinity of an equilibrium (DeAngelis 1975; de Ruiter et al. 1995; Brose et al. 2006; Neutel et al. 2007), is that the structure of food webs plays a key role in their stability at both the population and ecosystem levels. What is perhaps less clear at present is which specific structural properties of food webs explain their stability at the population and ecosystem levels and how they interact.

Last, an important limitation of most of the existing theory on the stability of communities, food webs, and ecosystems, including the new theory presented in this chapter, is that it deals with systems whose component species have fixed traits. Yet many species are able to adjust their behaviors, their life-history traits, and even their morphology to changes in their environment, in particular, to the presence or absence of their resources and predators. Such adaptive changes occur on time scales that range from very fast for behavioral adjustments to relatively slow or very slow for evolutionary adjustments (Abrams 1995). Incorporating the adaptive dynamics of species traits in food-web models is a major challenge that might alter the conclusions derived from models with fixed traits. Thus, Kondoh (2003) recently showed that fast enough adaptive changes in consumer food choice could turn a negative relationship between the complexity and stability of food webs into a positive one.

CONCLUSION

A striking feature of recent theoretical developments on the diversity–stability relationship is that they seek more precision and more biological realism than did previous, more abstract approaches. The most interesting aspect of recent theory within the context of this book is that it explicitly links stability properties at the population level and at the aggregate community or ecosystem level. By doing so, it reconciles previous theoretical results on the destabilizing effect of species diversity, which focused implicitly on population-level stability, and empirical observations that diverse ecosystems are often stable. It also departs from previous approaches by abandoning the convenient but unrealistic assumptions that the environment is constant and that populations and ecosystems reach equilibrium, and by explicitly incorporating species responses to environmental fluctuations.

As it turns out, temporal niche differentiation arising from differences in species environmental responses is the main mechanism that explains the stabilizing effect of species diversity on ecosystem properties. As a result of this explicit consideration of environmental fluctuations, recent theory has also shifted focus from local stability properties of equilibrium systems (such as qualitative stability and resilience) to temporal variability as a measure of the stability of specific population- or ecosystem-level properties. Temporal variability has the advantage of being simple, easily measured empirically, and applicable to nonequilibrium systems, thereby strengthening the links between theoretical and experimental studies.

Recent theoretical developments emphasize the key role played in the stability–diversity relationship by biological attributes of species such as their environmental responses, their food niche width, and the constraints that arise from trade-offs. It is comforting that biology eventually receives the attention it deserves in the theory of complex ecological systems. It is also comforting that the new theory is reaching other disciplines as the insurance value of biodiversity in the provision of ecosystem services is being formally incorporated in ecological economics (Armsworth and Roughgarden 2003; Baumgärtner 2007).

APPENDIX 5A
EFFECT OF SPECIES RICHNESS ON COMMUNITY VARIANCE IN THE STATISTICAL MODEL

Taking the derivative of community variance with respect to species diversity in equation (5.8) yields

$$\frac{\partial \sigma_{N_T}^2}{\partial S} = cm^z S^{-z} \left\{ \overline{\rho_N}(2S - 1) + 1 - z\left[\overline{\rho_N}(S - 1) + 1\right] + \frac{\partial \overline{\rho_N}}{\partial S} S(S - 1)\right\}.$$

(5A.1)

Assuming that the average correlation coefficient, $\overline{\rho_N}$, does not change with diversity (which is possible only if $\overline{\rho_N} \geq 0$), this derivative is negative, and hence community variance decreases as species diversity increases, provided

$$z > \frac{\overline{\rho_N}(2S - 1) + 1}{\overline{\rho_N}(S - 1) + 1},$$

(5A.2)

or, equivalently,

$$\overline{\rho_N} < \frac{z - 1}{S(2 - z) + z - 1}.$$

(5A.3)

Thus, the scaling coefficient z must be sufficiently large and the average correlation coefficient must be sufficiently low. Note that when $z \leq 1$, the average correlation coefficient must be negative. In this case, however, the assumption that it is independent of species richness does not hold because it tends to zero as the number of species increases (Loreau and de Mazancourt 2008). It is possible to show that the derivative of community variance with respect to species richness is then positive. Therefore, $z > 1$ is required for species richness to stabilize total biomass.

Material Cycling and the Overall Functioning of Ecosystems

So far I have moved gradually from simpler to more complex systems, starting with single populations (chapter 1), then continuing with competitive systems that have multiple species but a single trophic level (chapters 2, 3, and 5), and finally expanding the scope to food webs and interaction webs with multiple species and multiple trophic levels (chapters 4 and 5). Now has come the time to consider the ecosystem as a whole, and the specific constraints that arise from its overall functioning.

An ecosystem represents the entire system of biotic and abiotic components that interact in a given location. As such, it includes a wide range of biological, physical, and chemical processes that connect organisms and their environment. Ecosystems have been approached from a variety of perspectives. Some approaches have focused on biotic interactions, in particular, consumer–resource interactions. Ecosystems are then looked at from the point of view of food webs or interaction webs. I have already considered this point of view in chapter 4. Ecosystem ecology, however, has generally focused on the overall functioning of ecosystems as distinct entities, in particular, on patterns of energy and material flows. Energy flows within ecosystems have traditionally received the most attention (Lindeman 1942; E. P. Odum 1953; H. T. Odum 1983) because energy is a universal requirement of all biological processes and is relatively easy to measure. From a theoretical perspective, however, energy flows within ecosystems do not offer major new questions and challenges compared with the food-web perspective. Energy is transferred between organisms via trophic interactions and is gradually dissipated through respiration along the food chain. As a consequence, energy flows through the ecosystem from its fixation by photosynthesis to its dissipation by heterotroph respiration, with virtually no energy recycled within the ecosystem (figure 6.1, left).

By contrast, material elements are heavily recycled within ecosystems. *Material cycling* is an inevitable consequence of energy flow in any physical

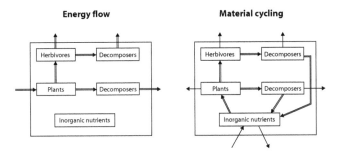

FIGURE 6.1. Contrasting effects of energy flow and material cycling in ecosystems. Energy (left) flows through the ecosystem, thus generating a linear chain of interactions, whereas material cycling (right) generates a circular causal chain that transmits feedbacks and indirect effects to all ecosystem components.

system at equilibrium when material exchanges across its boundaries are limited (Morowitz 1966). Without recycling of materials in limited supply, energy flow would stop rapidly, leading to ecosystem collapse. Biologically, material cycling is likely to emerge spontaneously through evolution of organisms that use other organisms or their waste products as resources. The Earth system as a whole, including the biosphere, the atmosphere, the hydrosphere, and the lithosphere, is virtually a closed system except for negligible inputs of trace elements via meteorites. That is why biogeochemical cycles play a critical role in the functioning of the Earth system. But even at the scale of local ecosystems, nutrients such as nitrogen and phosphorus generally limit plant growth, and hence the productivity of the ecosystem as a whole because primary production is the process on which the entire trophic pyramid is built. Accordingly, nutrient cycling is a key process in the overall functioning of local ecosystems. In most intact ecosystems, internal recycling accounts for the bulk of nitrogen and phosphorus taken up by organisms and released from organic matter each year—the amount of these elements that is recycled is typically an order of magnitude larger than the amount that enters or leaves terrestrial ecosystems (Vitousek and Matson 2009).

Material cycling has often been studied under its immediate aspect, that is, as a process by which the mineral elements necessary to primary production are renewed. But its main significance lies elsewhere, as the driver of a circular causal chain that transmits feedbacks and indirect effects to all ecosystem components (Ulanowicz 1990; Loreau 1998b) (figure 6.1, right). As such, it is a powerful organizing force of ecosystems, which imposes strong constraints on their overall functioning as well as on the dynamics and evolution of their component organisms.

The most fundamental and widespread interaction that lies at the core of material cycling in present-day ecosystems is that between autotrophs, or primary producers, and heterotrophs, or consumers. Autotrophs build high-energy organic compounds using external sources of energy and inorganic elements made available by heterotrophs, while heterotrophs dissipate energy and regenerate inorganic elements using the high-energy organic compounds produced by autotrophs. Lotka (1925) identified the special interest of this objective complementarity of functional roles in ecosystem functioning:

> "Coupled transformers are presented to us in profuse abundance, wherever one species feeds on another, so that the energy sink of the one is the energy source of the other.
>
> A compound transformer of this kind which is of very special interest is that composed of a plant species and an animal species feeding upon the former. The special virtue of this combination is as follows. The animal (catabiotic) species alone could not exist at all, since animals cannot anabolise inorganic food. The plant species alone, on the other hand, would have a very slow working cycle, because the decomposition of dead plant matter, and its reconstitution into CO_2, completing the cycle of its transformations, is very slow in the absence of animals, or at any rate very much slower than when the plant is consumed by animals and oxidized in their bodies. Thus the compound transformer (plant and animal) is very much more effective than the plant alone.
>
> [. . .] For it must be remembered that the output of each transformer is determined both by its mass and by its rate of revolution. Hence if the working substance, or any ingredient of the working substance of any of the subsidiary transformers, reaches its limits, a limit may at the same time be set for the performance of the great transformer as a whole. Conversely, if any one of the subsidiary transformers develops new activity, either by acquiring new resources of working substance, or by accelerating its rate of revolution, the output of the entire system may be reflexly stimulated. (Lotka 1925, pp. 330, 334–335)

Since Lotka, this hypothesis has been surprisingly little investigated theoretically, although the role of animals and decomposers as accelerators of organic matter decomposition and nutrient cycling has been postulated in many empirical studies. In this chapter, I shall explore the main features of material cycles, focusing on the cycling of a single limiting nutrient such as nitrogen or phosphorus. Limitation of ecosystem processes by multiple elements may be more common than previously believed (Elser et al. 2007), but the theoretical study of multiple coupled elemental cycles requires stoichiometrically explicit models, which are more complex and raise specific challenges that I shall not address here. In particular, I shall examine the ecosystem- and community-level consequences of the two main interactions between autotrophs and heterotrophs that are almost universally present in ecosystems, i.e., plant–decomposer and plant–herbivore interactions. I shall show that material cycling has the potential to qualitatively change the nature of species interactions through transmission of indirect effects along the cycle. I shall also briefly explore how horizontal diversity within autotrophs and heterotrophs may affect ecosystem functioning through material cycling.

NUTRIENT CYCLING IN A MINIMAL ECOSYSTEM MODEL

To explore some of the basic functional consequences of material cycles in ecosystems, let us begin by revisiting the minimal ecosystem model (1.23) presented in chapter 1. This model involved only two ecosystem compartments, i.e., plants (with nutrient stock P) and a limiting inorganic nutrient (with stock N). The limiting nutrient was partly recycled within the ecosystem, but the ecosystem was open to nutrient inputs and outputs. The model was defined by the following pair of dynamical equations:

$$\frac{dN}{dt} = I - qN - f(N)\,P + (1 - \lambda)mP,$$

$$\frac{dP}{dt} = f(N)\,P - mP,$$

(6.1)

where I is the input of inorganic nutrient in the ecosystem, q is the rate at which the inorganic nutrient is lost from the ecosystem, λ is the fraction of nutrient that is lost from the ecosystem once released by plants before or during the decomposition process, $f(N)$ is the plant functional response, and m is the nutrient turnover rate in plants.

We know from chapter 1 that, at equilibrium, the inorganic nutrient stock, the plant nutrient stock (which is proportional to plant biomass), and net primary production (measured by plant nutrient uptake) are, respectively,

$$N^* = f^{-1}(m),$$

(6.2)

$$P^* = \frac{I - qN^*}{\lambda m},$$

(6.3)

$$\Phi_P^* = mP^* = \frac{I - qN^*}{\lambda}.$$

(6.4)

Since plants are assumed not to interfere with each other, they control the equilibrium stock of their resource [equation (6.2)], as in classical consumer–resource theory. Equilibrium primary production [equation (6.4)] is the product of the net supply of inorganic nutrient to plants [the numerator of the right-hand side of equation (6.4)] and the efficiency with which plants conserve the nutrient within the ecosystem (as measured by the inverse of λ). Equilibrium plant biomass [equation (6.3)] is equal to primary production divided by the nutrient turnover rate in plants.

The term $1/\lambda$ in equation (6.4), which corresponds to plants' nutrient conservation efficiency, can be given a more precise interpretation. The probability of a unit amount of nutrient released by plants being recycled and used again by plants in the next cycle is $1 - \lambda$. Therefore, the average number of times a unit amount of nutrient is cycled by plants within the ecosystem is

$$1 + (1 - \lambda) + (1 - \lambda)^2 + \cdots = \sum_{i=0}^{\infty}(1 - \lambda)^i = \frac{1}{\lambda}. \tag{6.5}$$

This number, however, is conditional on the nutrient not being lost in inorganic form since this source of nutrient loss is accounted for in the net supply of inorganic nutrient in equation (6.4). Thus, *equilibrium primary production is equal to the net supply of inorganic nutrient to plants times the average number of times the nutrient is cycled by plants.*

As the above equations show, nutrient stocks and production are determined by simple, easily interpretable rules in ecosystems at equilibrium. These rules extend to more complex, multitrophic ecosystems, as I shall show below. Perhaps counterintuitively, equilibrium primary production is governed by the balance between nutrient inputs and outputs at the ecosystem level, not (or at least not directly) by internal process rates such as the decomposition rate of dead organic matter or the nutrient recycling rate. This conclusion seems to contradict Lotka's claim that "the output of each transformer is determined both by its mass and by its rate of revolution," as well as the deeply ingrained idea in ecology that consumers and decomposers have beneficial effects on ecosystem functioning by speeding up the decomposition process. We shall discuss the role of heterotrophs further in this chapter, but, for the time being, let us try to understand why this apparent contradiction occurs.

First, note that Lotka's statement is not formally contradicted by the theory developed here since equilibrium primary production is equal to the mass of nutrient, P^*, times its rate of revolution in plants, m [equation (6.4)]. This statement is trivially true since it describes the necessary relationship between fluxes and stocks in any physical system at equilibrium. But it does not provide any insight into the key mechanisms and parameters that determine primary production in ecosystems.

Second, Lotka's claim is valid even from a mechanistic viewpoint in a very special class of systems, i.e., closed ecosystems. A closed ecosystem is formally a limiting case of a chemostat (i.e., a system in which all compartments experience the same loss rate due to physical dilution) in which the dilution rate tends to zero (Loreau and Holt 2004). This limiting case can be recovered from model (6.1) using the following parameter transformations: (1) $I = qQ$, where q is the dilution rate and Q is the inflowing quantity of nutrient; (2) $m = q + r$, where r is the rate at which nutrient is released and recycled from plants; (3) $\lambda = q / (q + r)$; and (4) $q \rightarrow 0$. Equation (6.4) then reduces to

$$\Phi_p^* = (Q - N^*)r, \qquad (6.6)$$

and primary production is equal to the mass of nutrient locked up in biomass, $Q - N^*$, times its rate of revolution in biomass, r, as predicted by Lotka.

Given the importance played historically by closed systems in physics, it is not surprising that concepts and approaches borrowed from physics for closed systems have permeated ecology, especially in its early days. Since virtually all ecosystems are open to material exchanges, however, equation (6.6) is unlikely to apply to real ecosystems, except perhaps to the entire Earth system. Equation (6.4) is much more general and includes equation (6.6) as a special case. In open ecosystems, any equilibrium requires that outputs balance inputs, hence it is logical that the key parameters that determine material fluxes in such systems are those that govern their input–output balance.

Third, while process rates do not govern production at equilibrium, they do play a role in transient dynamics. The above equilibrium analysis is relevant only if the ecosystem can reach the equilibrium in a reasonable time period. When an ecosystem compartment has a very slow turnover rate, however, as is the case for instance with resistant soil organic matter in terrestrial ecosystems, this compartment can be considered constant on shorter time scales, and an equilibrium analysis can then be performed on the simplified system (de Mazancourt et al. 1998).

Last, although equilibrium primary production is not directly governed by internal process rates in open ecosystems, it is affected by these process rates indirectly, through their influence on the parameters and variables that determine the balance between nutrient inputs and outputs at the ecosystem level, such as the equilibrium inorganic nutrient stock and the fraction of nutrient lost by plants. For instance, if faster organic matter decomposition and nutrient recycling are accompanied by a smaller proportional loss of nutrient from the ecosystem, these increased process rates will indirectly affect primary production. This is why the nutrient recycling rate, r, is predicted to affect primary production in a closed ecosystem [equation (6.6)]. As mentioned above, a closed ecosystem is but a limiting case of a chemostat. In a chemostat, as r increases, the fraction of nutrient lost by plants through dilution, $\lambda = q / (q + r)$, decreases because the dilution rate q is kept constant. In most natural ecosystems, however, there is no necessary relationship between faster nutrient recycling and a smaller proportional loss of nutrient. In some cases, faster nutrient recycling can even lead to increased nutrient loss due to leaching or volatilization. While the rule encapsulated in equation (6.4) is general, its application to real ecosystems requires detailed knowledge of the factors that determine the relevant parameters in each ecosystem. In the rest of this chapter, I shall discuss a number of ways in which biological processes and species traits affect the nutrient input–output balance at the ecosystem level, and hence equilibrium primary and secondary production. Thus, biological processes and species traits do matter, but they matter for production at equilibrium to the extent that they affect nutrient inputs and outputs.

PLANT–DECOMPOSER INTERACTIONS: INDIRECT
MUTUALISM THROUGH NUTRIENT CYCLING

All ecosystems of the world, from the poorest ones in the most extreme environments to the richest ones in the most favorable environments, are characterized by a material cycle that involves at least two main partners: (1) plants or other autotrophs, which capture energy and inorganic nutrients to produce organic matter; and (2) heterotrophic decomposers, which consume organic matter and release nutrients in inorganic form. The reciprocal plant–decomposer interaction is donor-controlled on both sides; i.e., the consumption of plant material by decomposers is controlled by the amount of dead organic matter made available by plants and does not involve control of plant biomass by decomposers; similarly, the uptake of inorganic nutrient by plants is controlled by the amount of inorganic nutrient made available by decomposers and usually does not involve control of decomposer biomass by plants. As a matter of fact, the interaction is indirect since it is mediated by the abiotic pools of dead organic matter and inorganic nutrient.

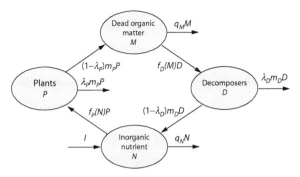

FIGURE 6.2. A simple ecosystem model that captures the main features of a material cycle involving autotrophic plants and heterotrophic decomposers. Circles represent nutrient stocks, while arrows represent nutrient fluxes. Modified from Loreau (1998b).

To explore the nature and functional consequences of the material cycle that results from this interaction, I present a simple, general ecosystem model that captures its essence and extends the previous model without decomposers (Loreau 1998b) (figure 6.2). As before, the ecosystem is assumed to be limited by a single nutrient; accordingly, all compartment sizes and fluxes correspond to nutrient stocks and fluxes. The inorganic nutrient pool (of size N) is supplied by a constant independent input I of inorganic nutrient per unit time. Plants produce dead organic matter, of which only the part (with nutrient stock M) that is readily accessible to decomposers is represented in the model. Recalcitrant dead organic matter is assumed for simplicity to be lost from the material cycle. It may either become unavailable and accumulate in the soil or be recycled on much longer time scales, in which case it is included in the constant input I. Plants and decomposers have nutrient stocks P and D, respectively. Their resource uptake depends on their respective stocks and functional responses to resource availability. The latter are represented by the functions $f_P(N)$ and $f_D(M)$, which may have any form provided that they are monotonic increasing. The fraction of nonassimilated nutrient consumed by decomposers returns to the dead organic matter compartment and therefore is not considered explicitly. Plants and decomposers release nutrient as a result of basal metabolism and mortality at rates m_P and m_D per unit time, respectively; these are equal to the turnover rates of plants and decomposers at equilibrium. A fraction λ_P or λ_D of these flows is lost from the system, the rest ($1 - \lambda_P$ or $1 - \lambda_D$) being cycled within the system in the form of readily available dead organic matter or inorganic nutrient. Nutrient is also lost from the pools of inorganic nutrient and dead organic matter (by leaching, export, etc.) at rates q_N and q_M per unit time, respectively.

Applying the principle of mass conservation and setting the time derivative of compartment size equal to the sum of inflows minus the sum of outflows for each compartment, we obtain the set of dynamical equations:

$$\frac{dN}{dt} = I - q_N N - f_P(N)P + (1 - \lambda_D)m_D D,$$

$$\frac{dP}{dt} = f_P(N)P - m_P P,$$

$$\frac{dM}{dt} = (1 - \lambda_P)m_P P - q_M M - f_D(M)D, \tag{6.7}$$

$$\frac{dD}{dt} = f_D(M)D - m_D D.$$

At equilibrium, the time derivatives in these equations vanish. Solving the resulting mass-balance equations provides the equilibrium nutrient stocks

$$N^* = f_P^{-1}(m_P),$$

$$P^* = \frac{S_N^*}{m_P \Lambda},$$

$$M^* = f_D^{-1}(m_D), \tag{6.8}$$

$$D^* = \frac{S_M^*}{m_D \Lambda},$$

where

$$S_N^* = I - q_N N^* - (1 - \lambda_D) q_M M^*,$$

$$S_M^* = (1 - \lambda_P)(I - q_N N^*) - q_M M^*, \tag{6.9}$$

$$\Lambda = \lambda_P + (1 - \lambda_P)\lambda_D.$$

These aggregate parameters are simple extensions of those defined for the model without decomposers. S_N^* is the excess of inflow of inorganic nutrient over outflows from the nonliving compartments, where the loss from dead organic matter is multiplied by the fraction $1 - \lambda_D$ of nutrient that is cycled by the decomposers; therefore, it represents the net supply of nutrient in inorganic form at equilibrium. S_M^* is interpreted similarly as the net supply of nutrient in the form of dead organic matter, while Λ represents the fraction of nutrient lost from the living compartments over a complete cycle. The equilibrium can be shown to always satisfy the Routh-Hurwitz criteria for local stability, hence to be qualitatively stable (May 1973; Puccia and Levins 1985).

The equilibrium primary (plant) production, Φ_P^*, and secondary (decomposer) production, Φ_D^*, are measured as before by the corresponding nutrient inflows, i.e.,

$$\Phi_P^* = f_P(N^*)P^* = \frac{S_N^*}{\Lambda},$$
$$\Phi_D^* = f_D(M^*)D^* = \frac{S_M^*}{\Lambda}. \tag{6.10}$$

Note that equilibrium primary and secondary productions have the same form as in the model without decomposers [equation (6.4)]. Equilibrium primary production is equal to the net supply of inorganic nutrient to plants, S_N^*, times the average number of times the nutrient is cycled by the living compartments, $1/\Lambda$. Similarly, equilibrium secondary production is equal to the net supply of organic nutrient to decomposers, S_M^*, times the average number of times the nutrient is cycled by the living compartments, $1/\Lambda$. Here again, the parameters that govern the nutrient input–output balance at the scale of the whole ecosystem determine nutrient fluxes and productions. Internal process rates come into play indirectly only insofar as they affect these key parameters.

In the limiting case of a closed ecosystem, we also obtain the same form as in the model without decomposers, for the same reasons. This case can be recovered from model (6.7) using the following parameter transformations: (1) $q_N = q_M = q$, where q is the dilution rate; (2) $I = qQ$, where Q is the inflowing quantity of nutrient; (3) $m_P = q + r_P$ and $m_D = q + r_D$, where r_P and r_D are the rates at which nutrient is released and recycled from plants and decomposers, respectively; (4) $\lambda_P = q / (q + r_P)$ and $\lambda_D = q / (q + r_D)$; and (5) $q \to 0$. Equations (6.10) then both reduce to

$$\Phi_P^* = \Phi_D^* = (Q - N^* - M^*)R, \tag{6.11}$$

where $R = (1/r_P + 1/r_D)^{-1}$. Since $1/r_P$ is the mean residence time of nutrient in plant biomass and $1/r_D$ is the mean residence time of nutrient in decomposer biomass, their sum is the mean residence time of nutrient in the living compartments over a complete cycle, and R is the rate of revolution of nutrient in the living compartments over a complete cycle. Thus, equilibrium primary and secondary productions are equal to the mass of nutrient locked up in biomass, $Q - N^* - M^*$, times its rate of revolution in biomass, R, just as before.

Although the similarity between the rules that govern ecosystems with and without decomposers is striking, an additional insight that emerges from the model including decomposers is that *the productions of the various living compartments are coupled in a material cycle* (they are even identical in the case of a closed ecosystem). Anything that affects one component of an ecosystem simultaneously affects all the other components of that ecosystem. In the case of the plant–decomposer system, any parameter that has a positive effect on primary production also has a positive effect on secondary production, and vice versa. Thus, material cycling generates an indirect mutualism between the two partners. Note, however, that this conclusion hinges on there being a single limiting nutrient. When several nutrients are potentially limiting, differences in the stoichiometric requirements of the two partners can generate competition between plants and decomposers for the same nutrient under some conditions, and even ecosystem collapse under extreme conditions (Daufresne and Loreau 2001; Cherif and Loreau 2007).

Two sets of critical parameters that are under the control of species traits and that affect the ecosystem-level nutrient input–output balance are the resource competitive abilities and nutrient cycling efficiencies of plants and decomposers. Classical resource competition theory (chapter 2) states that the competitive ability of either plants or decomposers is determined by their ability to deplete their respective resources, i.e., by their resource-use intensity. The plant species with the lowest N^* will displace all other plant species; similarly the decomposer species with the lowest M^* will displace all the others. Thus, competitive ability may be measured conveniently by the inverse of N^* or M^*. As the competitive ability of either plants or decomposers increases as a result of interspecific competition, the nutrient losses from the abiotic compartment they control decrease, and hence, by equations (6.9), the corresponding net nutrient supplies increase. As a consequence, ecosystem cycling efficiency, as measured by the recycled fraction of inflows to the nutrient inorganic pool, or, equivalently, by the probability that a molecule of nutrient completes a full cycle (Finn 1980), increases. By equations (6.10), primary production and secondary production also increase (figure 6.3).

Equations (6.10) also predict that species traits that improve the nutrient cycling efficiency of either plants or decomposers (i.e., that decrease either λ_P or λ_D) should have a strong positive effect on ecosystem cycling efficiency, primary production, and secondary production (figure 6.4). There are many examples of traits that may play this role in plants. For instance, plants may produce litters of different qualities, thereby controlling patterns of nutrient cycling (Hobbie 1992); they may modify soil structure, which in turn strongly affects nutrient retention (Wood 1984); they may recycle some limiting nutrients internally via biochemical pathways (Switzer and Nelson 1972); or they may directly control nitrification, and hence

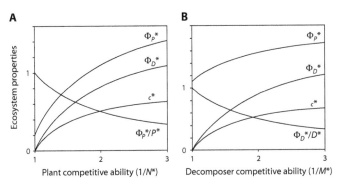

FIGURE 6.3. Equilibrium primary productivity ($\Phi_P{}^*$), secondary productivity ($\Phi_D{}^*$), plant productivity/biomass ratio ($\Phi_P{}^*/P^*$), decomposer productivity/ biomass ratio ($\Phi_D{}^*/D^*$), and ecosystem cycling efficiency (c^*) as functions of the competitive ability of either plants ($1/N^*$, A) or decomposers ($1/M^*$, B) in the model depicted by figure 6.2. In both cases, the productivity and biomass of both plants and decomposers increase; the productivity/biomass ratio of either plants (in A) or decomposers (in B) decreases, while the other is constant; and ecosystem cycling efficiency increases. Ecosystem cycling efficiency (multiplied by 2 on the graphs) is measured by the recycled fraction of inflows to the nutrient inorganic pool. Because there is a single cycle in this system, this measure can be shown to be equivalent to the probability that a molecule of nutrient completes a full cycle, and to Finn's (1980) cycling index. Lotka–Volterra functions were used for resource uptakes, the corresponding functional responses being $g_P(N) = c_P N$ and $g_D(M) = c_D M$. Only the nutrient turnover rates m_P (in A) and m_D (in B) were varied; the other parameters were set at the following values: $I = 1.2$; $c_P = c_D = q_N = q_M = 1$; $\lambda_P = 0.1$; $\lambda_D = 0.5$; $m_D = 0.18$ in A; $m_P = 0.0889$ in B. Modified from Loreau (1998b).

nitrogen outputs, in the vicinity of their rooting system through inhibition of nitrifying bacteria (Lata et al. 2004).

In all these cases, an *indirect mutualistic interaction* emerges between plants and decomposers, mediated by nutrient cycling. The production and biomass of each partner are boosted by any trait change that also increases the production of the other partner.

PLANT–HERBIVORE INTERACTIONS: FROM DIRECT EXPLOITATION TO INDIRECT MUTUALISM

The insight that decomposers have an indirect mutualistic relationship with plants through material cycling is important, but it does not go against common sense because decomposers do not consume plants directly. Much

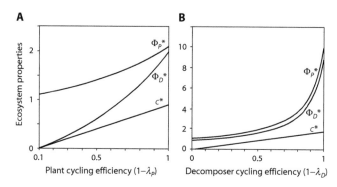

FIGURE 6.4. Equilibrium primary productivity (Φ_P^*), secondary productivity (Φ_D^*), and ecosystem cycling efficiency (c^*) as functions of the nutrient cycling efficiency of plants ($1 - \lambda_P$, A) or decomposers ($1 - \lambda_D$, B) in the model depicted by figure 6.2. In both cases, the productivity and biomass of both plants and decomposers, as well as ecosystem cycling efficiency, increase when nutrient cycling efficiency increases. Productivity/biomass ratios are not affected. Ecosystem cycling efficiency is multiplied by 2 on the graphs. Parameters values are as follows: $I = 1.2$; $q_N N^* = q_M M^* = 0.1$; $\lambda_D = 0.5$ in A; $\lambda_P = 0.1$ in B. Modified from Loreau (1998b).

less intuitive is the fact that herbivores too can have a mutualistic relationship with plants despite the fact that they consume them. The idea that consumers are detrimental to their food resources is deeply engraved on our civilization. The need for a smooth functioning of the economy imposes a constant fight against other animal species feeding on our plant food resources, which are therefore viewed as undesirable pests from which we must protect ourselves. Ecology as a science seeks to establish a more balanced view of nature. Even in ecology, however, plant–herbivore interactions have been regarded as essentially antagonistic because herbivores have an obvious negative direct effect on plants through biomass consumption.

This traditional view has been challenged by the *grazing optimization hypothesis*, which states that primary productivity, or even plant fitness, is maximized at an intermediate rate of herbivory (Owen and Wiegert 1976, 1981; McNaughton 1979; Hilbert et al. 1981). This hypothesis is supported by some empirical data, for instance, from the Serengeti savanna ecosystem (figure 6.5), but it has been strongly controversial (Silvertown 1982; Belsky 1986; McNaughton 1986; Belsky et al. 1993; Lennartsson et al. 1997). Several mechanisms are likely to generate increased primary production following grazing. Some of these mechanisms are physiological and operate in the short term. For instance, plant growth can be stimulated by removal of apical dominance or mobilization of stored resources after a grazing

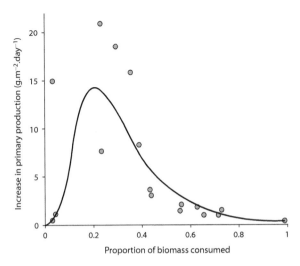

FIGURE 6.5. Grazing optimization in the Serengeti: aboveground primary production is increased compared with its background level in the absence of grazers and peaks at an intermediate grazing intensity (proportion of plant biomass consumed by grazers). Dots are data points, and the curve is a non-linear model fitted to data ($r^2 = 0.69$, $P < 0.001$). Modified from McNaughton (1979).

event. The only mechanism likely to generate and maintain grazing optimization in the long term, however, is nutrient cycling because of the sustained positive indirect effects it transmits to all ecosystem components.

Claire de Mazancourt and I have devoted a series of studies to identifying and predicting the general conditions under which nutrient cycling is expected to generate grazing optimization (Loreau 1995; de Mazancourt et al. 1998, 1999; de Mazancourt and Loreau 2000a). Since our conclusions are rather robust, here I shall only present a simple ecosystem model to derive these conditions and discuss their implications.

Assume, as before, that the ecosystem is limited by a single nutrient. All compartment sizes and fluxes are then measured by the corresponding nutrient stocks and fluxes. Both plants (P) and herbivores (H) produce dead organic matter, but the physical and chemical properties of plant detritus (M_P) and herbivore detritus (M_H) are different, leading to two distinct nutrient recycling pathways (figure 6.6). The fact that some decomposers may be involved in the two recycling pathways simultaneously is irrelevant here for, as we shall see, the overall efficiency of each recycling pathway is what counts. Therefore, decomposers are ignored altogether here. Including them explicitly does not alter any of the results presented

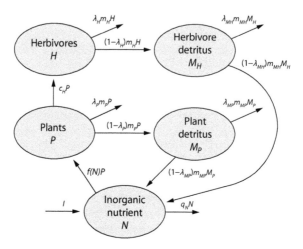

FIGURE 6.6. A simple ecosystem model to analyze the indirect effects of herbivores on plants through nutrient cycling. Circles represent nutrient stocks, while arrows represent nutrient fluxes.

below (de Mazancourt et al. 1999). Inorganic nutrient (N) is supplied by a constant input I and lost at a rate q_N per unit mass. Plants take up inorganic nutrient following a functional response $f(N)$ and release nutrient as dead organic matter at a rate m_P. For simplicity, herbivory is assumed to be a donor-controlled function proportional to plant biomass with a rate constant c_H. Traditional functional responses leading to recipient control are more complex to analyze but lead to identical results (de Mazancourt et al. 1998). Herbivores release nutrient as dead organic matter at a rate m_H. Egestion of nutrient consumed but not assimilated by herbivores is included in dead organic matter production. Plant detritus and herbivore detritus are then decomposed, releasing inorganic nutrient at rates m_{MP}, and m_{MH}, respectively. Nutrient, however, can be lost from any of the organic compartments. The fraction of nutrient lost from the ecosystem once released by any compartment X is denoted by λ_X.

As before, the dynamical equations corresponding to figure 6.6 are obtained easily by setting the rate of change of each compartment equal to the sum of inflows to that compartment minus the sum of outflows from that compartment. This system reaches a stable equilibrium for the following values of nutrient stocks:

$$N^* = f^{-1}(m_P + c_H),$$

$$P^* = \frac{S_N^*}{(m_P + c_H)\Lambda},$$

$$H^* = \frac{c_H P^*}{m_H}, \tag{6.12}$$

$$M_P^* = \frac{(1 - \lambda_P)m_P P^*}{m_{MP}},$$

$$M_H^* = \frac{(1 - \lambda_H)c_H P^*}{m_{MH}},$$

where

$$S_N^* = I - q_N N^*,$$

$$\Lambda = \frac{m_P \Lambda_P + c_H \Lambda_H}{m_P + c_H}, \tag{6.13}$$

$$\Lambda_P = 1 - (1 - \lambda_P)(1 - \lambda_{MP}),$$

$$\Lambda_H = 1 - (1 - \lambda_H)(1 - \lambda_{MH}).$$

At this equilibrium, primary (plant) production and secondary (herbivore) production are, respectively,

$$\Phi_P^* = \frac{S_N^*}{\Lambda},$$

$$\Phi_H^* = \left(\frac{c_H}{m_P + c_H}\right)\Phi_P^*. \tag{6.14}$$

These equations have straightforward interpretations. S_N^* is the net supply of inorganic nutrient to plants. Λ_P and Λ_H are the probabilities that a molecule of nutrient is lost from the ecosystem along the plant and herbivore recycling pathways, respectively. Therefore, Λ, which is a weighted average of these two probabilities, is the probability that a molecule of nutrient is lost during the recycling process as a whole. Thus, equilibrium primary production is equal to the net supply of inorganic nutrient to plants times the average number of times the nutrient is cycled by the living compartments and their dependent detritus, $1/\Lambda$, as in previous models. Equilibrium secondary production is here proportional to primary production, its proportion being determined by the proportional share of herbivory among the various factors that contribute to the total nutrient turnover rate in plants[1], $m_P + c_H$. Here again, the parameters that govern the nutrient input–output balance at the scale of the whole ecosystem determine nutrient fluxes and productions. It is straightforward to show that the interpretation of equilibrium primary production as a product of a mass of nutrient and its rate of revolution is valid only for closed ecosystems, as in previous models (Loreau 1995).

[1] In fact, to be correct, secondary production is smaller than this since it should be multiplied by herbivore assimilation efficiency to account for egestion of consumed plant material that is not assimilated.

This model also shows clearly that the number of compartments involved in the various nutrient recycling pathways does not affect the equilibrium properties of the system as long as the interactions between these compartments are donor-controlled. The equilibrium stocks of plant and herbivore detritus are proportional to those of plants and herbivores [equations (6.12)], and the parameters that determine equilibrium ecosystem properties are lumped nutrient loss probabilities along entire nutrient recycling pathways [equations (6.13) and (6.14)]. Therefore, plant and herbivore detritus could have been removed from the model without altering its equilibrium properties, provided the parameters in the simplified model were properly interpreted.

Since increasing grazing intensity, c_H, increases the equilibrium level of inorganic nutrient [equation (6.12)] and hence decreases the net supply of inorganic nutrient [equation (6.13)], plant biomass and primary production ultimately decrease to zero at very high grazing intensities. Therefore, grazing optimization occurs if and only if primary production increases for low values of grazing intensity, i.e., if and only if

$$\left(\frac{d\Phi_P^*}{dc_H}\right)_{c_H=0} > 0. \tag{6.15}$$

This condition becomes, after some algebraic manipulation,

$$\Lambda_H < \Lambda_P\left(1 - \frac{q_N \Delta N_0}{I - q_N N_0^*}\right), \tag{6.16}$$

where N_0^* is the equilibrium inorganic nutrient stock in the absence of herbivory ($c_H = 0$), and $\Delta N_0 = m_P/f'(N_0^*)$ is inversely proportional to the sensitivity of plant nutrient uptake to inorganic nutrient availability at equilibrium in the absence of herbivory (figure 6.7).

The term in parentheses in inequality (6.16) is < 1. Therefore, grazing optimization requires that the fraction of nutrient lost along the herbivore recycling pathway, Λ_H, be sufficiently smaller than the fraction of nutrient lost along the plant recycling pathway, Λ_P. In other words, *grazing optimization requires that herbivores improve the ecosystem's nutrient conservation efficiency enough to compensate for the loss of plant biomass to grazing.* Inequality (6.16), however, cannot be met if the term in parentheses is negative. Therefore, grazing optimization also requires that

$$I > q_N (N_0^* + \Delta N_0); \tag{6.17}$$

that is, the input of inorganic nutrient must be large enough to outweigh nutrient leaching in the absence of herbivores (a necessary condition for plant persistence) plus an additional term that depends on the sensitivity of plant nutrient uptake to inorganic nutrient availability.

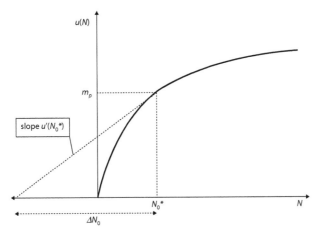

FIGURE 6.7. Plant nutrient uptake rate, $u(N)$, as a function of the amount of inorganic nutrient in the model depicted by figure 6.6. The slope of this curve, $u'(N)$, represents plant sensitivity to an increase in the amount of inorganic nutrient. The parameter $\Delta N_0 = m_p/u'(N_0^*)$, which appears in the condition for grazing optimization [inequalities (6.17) and (6.18)], is inversely proportional to the sensitivity of plant nutrient uptake to inorganic nutrient availability at the equilibrium without herbivores, N_0^*. The condition for grazing optimization is more easily fulfilled if plant nutrient uptake is highly sensitive to an increase in inorganic nutrient availability. Modified from de Mazancourt et al. (1998).

When these conditions are met, primary production reaches a maximum for an intermediate value of grazing intensity, generating a typical grazing optimization curve (figure 6.8). By contrast, plant biomass always decreases as grazing intensity increases. Heterotrophic consumers can boost primary production by enhancing the ecosystem's nutrient cycling efficiency, but they cannot boost plant biomass because they divert the increased primary production to their own benefit. Note that the intermediate value of grazing intensity at which primary production is maximized can sometimes be surprisingly high. When we applied our model to an African savanna ecosystem in Ivory Coast, for which we established a complete nitrogen budget based on field measurements, we predicted that, provided herbivores were efficient enough at recycling nitrogen, grazing optimization would occur at grazing intensities that correspond to up to 90 percent of primary production consumed (figure 6.9). The critical fraction of nitrogen lost along the herbivore recycling pathway, Λ_H, below which grazing optimization should occur in this savanna ecosystem, was estimated to be 0.24 on a time scale of decades, and 0.19 on a time scale of

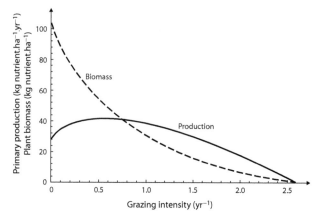

FIGURE 6.8. Equilibrium primary production and plant biomass vs. grazing intensity when the conditions for grazing optimization are met in the model depicted by figure 6.6. Modified from de Mazancourt et al. (1998).

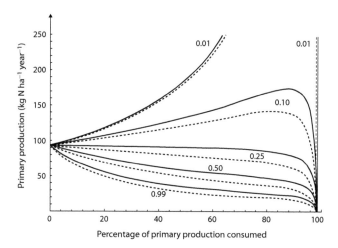

FIGURE 6.9. Equilibrium primary production versus percentage of primary production consumed by herbivores predicted by a model parameterized for the nitrogen cycle in a savanna ecosystem in Ivory Coast. Solid lines show predictions for a short-term equilibrium on a time scale of decades, while dashed lines show predictions for a long-term equilibrium on a time scale of centuries. The five pairs of curves correspond to different values of the fraction of nitrogen lost along the herbivore recycling pathway, Λ_H. The critical values of Λ_H below which grazing optimization occurs is 0.24 for the short-term equilibrium, and 0.19 for the long-term equilibrium. Modified from de Mazancourt et al. (1999).

centuries, the difference between these figures resulting from the fact that different ecosystem processes are expected to reach equilibrium on these time scales (de Mazancourt et al. 1999). Based on data collected in the literature, we concluded that herbivores should be efficient enough at recycling nitrogen to keep the fraction of nitrogen lost under these critical values and hence to increase primary production. The main mechanism that makes grazing optimization possible in this ecosystem is simple: a large fraction of the nitrogen stocked in grass biomass is lost to annual fires, while large herbivores contribute to keep it within the ecosystem by reducing grass biomass.

The conclusions reached here using the simple model above are robust to changes in model structure that increase the number of ecosystem compartments (de Mazancourt et al. 1999). Unfavorable indirect effects of herbivory, such as replacement of a productive plant species by a less productive one, affect these conclusions quantitatively, but grazing optimization is still possible even under these more restrictive conditions (de Mazancourt and Loreau 2000a). Thus, although herbivores have a negative direct effect on plants through consumption of plant tissue, they have a positive indirect effect through nutrient cycling. This positive indirect effect can outweigh the negative direct effect, leading potentially to an *indirect mutualistic interaction* between the two partners. The beneficial effects of herbivores on plants, however, concern plant production, not plant biomass. Therefore, the nature of the interaction from the plants' viewpoint depends on whether their fitness depends mostly on their production or on their biomass. I shall further discuss this issue in chapter 8.

HORIZONTAL DIVERSITY, NUTRIENT CYCLING, AND ECOSYSTEM FUNCTIONING

Previous chapters have considered the relationship between biodiversity and ecosystem functioning at length. Therefore, it is worth examining here how horizontal diversity may affect ecosystem functioning through material cycling. Very few theoretical studies have addressed this topic. Paradoxically, the first theoretical work on biodiversity and ecosystem functioning that I know of is my own pioneering work (Loreau 1996), which was based on a comprehensive multitrophic ecosystem model and highlighted the importance of nutrient dynamics and cycling for the relationship between biodiversity and ecosystem functioning. Although subsequent models were generally simpler and more specific, most of the results obtained using my early model can be understood in retrospect by combining the various principles pertaining to horizontal diversity, vertical diversity, and material cycling that I presented in chapters 3 and 4 and in the previous sections of this chapter.

One of the additional features that material cycling brings to the relationship between biodiversity and ecosystem functioning, however, is the potential for a diversity of nutrient cycling pathways in an ecosystem. How does this diversity of nutrient cycling pathways affect ecosystem functioning? In a microcosm experiment that manipulated the diversity of primary producers (algae) and decomposers (bacteria) simultaneously, Naeem et al. (2000) found complex interactive effects of algal and bacterial diversity on algal and bacterial biomass production. Both algal and bacterial diversity had significant effects on the number of carbon sources used by bacteria, suggesting nutrient cycling associated with microbial exploitation of organic carbon sources as the link between bacterial diversity and algal production. To examine this issue theoretically, I extended the plant–decomposer model (6.7) discussed above by including a diversity of plant organic compounds and a diversity of microbial species (Loreau 2001). Since producer diversity affects decomposers through the diversity of the organic compounds they produce, I did not represent plant diversity explicitly in the new model but only the diversity of their organic compounds (C_j), which collectively replace the litter compartment of the previous model (figure 6.10). The various organic compounds may differ in aspects of chemical quality such as their C:N ratio and, consequently, represent distinct resources for decomposers.

The dynamical equations corresponding to figure 6.10 are as follows:

$$\frac{dN}{dt} = I - q_N N - f_P(N)P + \sum_i \mu_{Di} m_{Di} D_i, \tag{6.18a}$$

$$\frac{dP}{dt} = f_P(N)P - m_P P, \tag{6.18b}$$

$$\frac{dC_j}{dt} = p_j \mu_P m_P P - m_{Cj} C_j, \tag{6.18c}$$

$$\frac{dD_i}{dt} = \sum_i \pi_{ji} \mu_{Cj} m_{Cj} C_j - m_{Di} D_i, \tag{6.18d}$$

where

$$\sum_j p_j = \sum_i \pi_{ji} = 1. \tag{6.19}$$

Most parameters are the same as in model (6.7), but the fraction of nutrient recycled within the ecosystem once released by any compartment X is here denoted by $\mu_X = 1 - \lambda_X$. Also, the consumption of organic compounds by

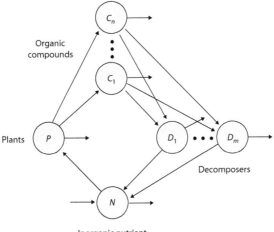

Organic
compounds

Plants

Decomposers

Inorganic nutrient

FIGURE 6.10. Flow diagram of the ecosystem model used to analyze the inter-active effects of plant organic compound diversity and microbial diversity on ecosystem functioning. Circles represent nutrient stocks, while arrows repre-sent nutrient fluxes. Modified from Loreau (2001).

decomposers is assumed to be donor-controlled for the sake of simplicity. The analysis is more complicated with recipient-controlled interactions, but the results are qualitatively similar (Loreau 2001). Last, p_j is the frac-tion of nutrient recycled from primary producers that is stored in the form of organic compound j, and π_{ji} is the fraction of nutrient recycled from or-ganic compound j that is used by decomposer species i. To further simplify the analysis and focus on the specific effects of organic compound diversity and microbial diversity, I shall further assume that decomposers are equiv-alent except for the way they use resources; i.e., $\mu_{Di} = \mu_D$ and $m_{Di} = m_D$ for all species.

In this case, primary (plant) productivity, Φ_P^*, plant biomass, P^*, sec-ondary (microbial) productivity, Φ_D^*, and decomposer biomass, D^*, at equilibrium take on following form:

$$\Phi_P^* = \frac{S_N^*}{\Lambda},\tag{6.20a}$$

$$P^* = \frac{\Phi_P^*}{m_P},\tag{6.20b}$$

$$\Phi_D^* = \frac{\mu_P \overline{\mu_C} S_N^*}{\Lambda},\tag{6.20c}$$

$$D^* = \frac{\Phi_D^*}{m_D},\tag{6.20d}$$

where

$$S_N^* = I - q_N N^*, \tag{6.21a}$$

$$\Lambda = 1 - \mu_P \mu_D \overline{\mu_C}, \tag{6.21b}$$

$$\overline{\mu_C} = \sum_j p_j \mu_{Cj}. \tag{6.21c}$$

In these equations, S_N^* is the net supply of nutrient in inorganic form at equilibrium, Λ is the fraction of nutrient lost from the organic compartments over a complete nutrient cycle, and $\overline{\mu_C}$ is the average fraction of nutrient recycled from organic compounds to decomposers, a measure of nutrient recycling efficiency from organic compounds to decomposers.

It is apparent from these equations that ecosystem properties depend on the diversity of organic compounds and on the diversity of decomposers through the single aggregate parameter $\overline{\mu_C}$. Increasing nutrient recycling efficiency within the ecosystem, or, equivalently, reducing the average proportion of nutrient lost to the ecosystem, contributes to increasing both primary and secondary productivity and producer and decomposer biomass. To explore in more detail how horizontal diversity affects ecosystem processes through nutrient recycling efficiency, assume that the limiting nutrient is released from organic compound j through two independent pathways: it may either leave the system at a rate l_{Cj} or be consumed by decomposer i at a rate c_{ji}. In this case,

$$m_{Cj} = l_{Cj} + \sum_i c_{ji}, \tag{6.22a}$$

$$\mu_{Cj} = \frac{\sum_i c_{ji}}{l_{Cj} + \sum_i c_{ji}}. \tag{6.22b}$$

The outcome now depends on the c_{ji}, which determine the decomposer niche height (absolute resource-use intensity), niche breadth (degree of generalization), and niche overlap (resource-use similarity between species). I examined four simple scenarios for decomposer niches in which all species have identical niche breadths (same number of compounds used) and identical niche heights (same total consumption rates, except in the fourth scenario):

1. All species are specialized on the same organic compound k (the SSC scenario): for all i, $c_{ji} = c'$ for $j = k$ and $c_{ji} = 0$ for $j \neq k$.
2. All species are specialized on different organic compounds (the SDC scenario): for all i, $c_{ji} = c'$ for $j = i$ and $c_{ji} = 0$ for $j \neq i$.

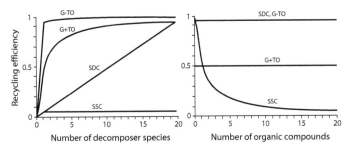

FIGURE 6.11. Effects of microbial diversity, m (left), and plant organic compound diversity, n (right), on nutrient recycling efficiency from organic compounds to decomposers, $\overline{\mu_C}$, for the SSC (specialists on the same compound), SDC (specialists on different compounds), G+TO (generalists with trade-off), and G–TO (generalists without trade-off) scenarios in the model depicted by figure 6.10. In left panel, $p_j = 1/n$ and $l_{C_j} = c'/n$ for all compounds j, and $m \leq n = 20$. In right panel, $p_j = 1/n$ and $l_{C_j} = c'/m$ for all compounds j, and $n \leq m = 20$. Modified from Loreau (2001).

3. All species are complete generalists using all organic compounds, but there is a trade-off between their degree of generalization and their ability to consume each compound (the G+TO scenario): $c_{ji} = c'/n$ for all i and j.
4. All species are complete generalists using all organic compounds, and there is no trade-off between generalization and consumption of each compound (the G–TO scenario): $c_{ji} = c'$ for all i and j.

The first two and last two scenarios are limiting cases of minimum and maximum niche breadth, respectively; SSC and SDC are limiting cases of maximum and minimum niche overlap, respectively; and G+TO and G–TO contrast scenarios with different niche heights. Niche breadth may be interpreted concretely as the diversity of enzymes that allows a microbial species to break down a diversity of organic compounds with different C:N ratios, while niche height may be interpreted as a species' potential enzymatic activity.

Figure 6.11 (left) shows that microbial diversity always has a positive effect on nutrient recycling efficiency from organic compounds to decomposers and hence on all ecosystem properties. The magnitude of this effect, however, is strongly dependent upon microbial niche breadth and overlap: it is usually smallest under SSC because all species are specialized on the same compound so that recycling efficiency cannot exceed the proportion of this compound in total compound production (here, 5 percent), and largest under SDC because microbial diversity decreases the number of unused compounds, thus improving nutrient conservation in direct proportion to the number of species. The effect of microbial diversity quickly saturates when decomposers are generalists, and more quickly so when resource consumption is higher (under G–TO as compared with G+TO).

In contrast, as long as microbial diversity or generalization is high enough that all compounds can be used, organic compound diversity has no effect on nutrient conservation and ecosystem processes (SDC, G+TO, and G–TO in figure 6.11, right). When increasing organic compound diversity means increasing the number of compounds that cannot be used by decomposers (SSC), it has a negative effect on ecosystem processes because these compounds are effectively lost from the material cycle at the time scale considered.

Thus, the model predicts that microbial diversity has a positive effect on nutrient recycling efficiency and ecosystem processes through either greater intensity of microbial exploitation of organic compounds or functional niche complementarity, just as at other trophic levels (chapter 3). Microbial niche breadth and overlap affect ecosystem processes only if they increase the number of organic compounds that are decomposed. In contrast, plant organic compound diversity has a negative effect, or at best no effect, on ecosystem processes because it contributes to generate unused resources that decrease overall nutrient cycling efficiency. This prediction agrees with experiments that showed no consistent effects of plant litter diversity on litter decomposition (e.g., Hector et al. 2000). The combination of the potentially opposite effects of plant litter diversity and microbial diversity on ecosystem functioning might explain Naeem et al.'s (2000) experimental results.

Hättenschwiler and Gasser (2005), however, found significant interactive effects of litter diversity and soil fauna on litter decomposition in a temperate forest: litter decomposition was enhanced by litter diversity in the presence of specific groups of detritivores. This finding suggests that litter produced by different plant species might have synergistic effects on its consumption by some detritivores. Similarly, some microbes might work in consortia to break down organic compounds, thus generating synergistic effects of microbial diversity. These synergistic effects were not taken into account in the above model and have the potential both to strengthen the positive effect of microbial diversity on ecosystem functioning and to turn the negative effect of plant litter diversity into a positive one. Building a theory of synergistic interactions between alternative nutrient recycling pathways is an exciting challenge for future research. But more empirical data are needed to support such a theory in order to make it robust and relevant.

MATERIAL CYCLING AND ECOSYSTEM DEVELOPMENT

Plant and animal communities show orderly changes in species composition and community structure as they colonize newly created biotopes or after major disturbances. This regular sequence of species and of associated patterns is encapsulated in the concept of *ecological succession*. Although ecological successions have a strong stochastic component, they are often remarkably predictable in the long run. This allows, for instance, successions of necrophagous insects on corpses to be used in legal medicine to estimate the time of death. The study of ecological successions is a classical topic in community ecology. Succession theories typically address the relative importance of different mechanisms that govern the gradual replacement of species as succession proceeds (Connell and Slatyer 1977; Walker and Chapin 1987).

In a seminal paper, E. P. Odum (1969) extended the scope of studies on ecological succession to patterns of changes in ecosystem properties as ecosystems mature, which led him to introduce the broader concept of *ecosystem development*. This integrative concept included predictable changes not only in patterns of energy and material flows but also in species life-history traits, species diversity, food-web structure, and other community properties. In this section I shall concentrate more specifically on changes in ecosystem properties such as biomass, production, and nutrient cycling efficiency and examine some of the consequences of the theory of material cycling that I have expounded above for ecosystem development.

Plant successions occur on time scales that range typically from years to centuries. Ecosystem development on these time scales shows some rather general trends in ecosystem properties, which include increased biomass (in particular, plant biomass), increased production (in particular, net primary production), and decreased productivity per unit biomass (e.g., Begon et al. 1996). E. P. Odum (1969) also postulated that nutrient cycling should become more efficient, leading to more closed material cycles as succession proceeds. A number of phenomenological holistic theories based on thermodynamic principles have been proposed to account for these trends during succession (H. T. Odum 1983; Ulanowicz and Hannon 1987; Schneider and Kay 1994; Fath et al. 2004). Although attractive, these theories lack a mechanistic basis in ecology and even more so in evolution, as I shall discuss further in chapter 8. The simple material cycle model that I presented earlier to explore the functional consequences of plant–decomposer interactions [equations (6.7)] yields powerful predictions about changes in ecosystem properties during succession.

Ecological succession is a complex phenomenon that includes several processes. One of its most salient features, which gave birth to the term "succession" itself, is the progressive replacement of species with high colonization ability and growth rate by species with greater size and life span (E. P. Odum 1969; Tilman 1988). These factors contribute to decrease biomass turnover rate and hence increase resource-use intensity. Since the plant functional response $f_p(N)$ is a monotonic increasing function of N, $N^* = f_p^{-1}(m_p)$ is an increasing function of the turnover rate m_p, all other things being equal. A similar argument holds for decomposers, though likely on a shorter time scale. Provided that the colonization abilities and growth rates of the successive species are different enough, succession may be approximated by a shifting trajectory of steady-state communities This leads to a simple prediction: species replacement during succession makes ecosystems develop toward increasing competitive ability, that is, to move to the right along the x-axes in figure 6.3. Tilman (1988) and Tilman and Wedin (1991) provided strong theoretical and experimental evidence for an increase in plant resource-use intensity during succession. This directional change in species life-history traits is the mechanistic basis of the classical competition–colonization trade-off (chapter 2). As a direct consequence of this change, both primary and secondary productivities are expected to increase during succession (figure 6.3). Because the productivity/biomass ratio of a compartment is simply equal to its turnover rate at steady state, this ratio tends to decrease, which means that biomass not only increases but also increases faster than productivity. Finally, the ecosystem's cycling efficiency also increases because plants and decomposers utilize their inorganic and organic resources more efficiently, thereby reducing direct losses of these resources.

Other species traits that could generate similar successional trends in ecosystem properties are those that control the nutrient cycling efficiency of either plants or decomposers. If late-successional plants, for instance, were generally better able than early-successional plants to keep the limiting nutrient within the ecosystem through any of the mechanisms discussed earlier, as E. P. Odum (1969) hypothesized, this would contribute to strongly enhance ecosystem cycling efficiency, primary production, and secondary production (figure 6.4). Unfortunately, the empirical and experimental evidence for this hypothesis is still scarce.

In conclusion, the dynamics of species replacement driven by resource competition within material cycles is able to explain all the major trends in the basic functional properties of ecosystems during their development. This conclusion applies a fortiori to ecosystem development on an evolutionary time scale since in that case transient dynamics can be ignored safely and new genotypes can invade only if they have a greater competitive ability. I shall discuss more about the evolutionary dimension of ecosystem development in chapter 8.

At first sight some of these conclusions seem to contradict other patterns observed during succession. In particular, Vitousek and Reiners (1975) reasoned that the gradual buildup of plant biomass during succession should be accompanied by decreased losses of the limiting nutrient from the ecosystem because plants accumulate the nutrient, thereby reducing nutrient outputs below nutrient inputs. But as succession proceeds, nutrient outputs must eventually increase to balance nutrient inputs. Thus, nutrient losses are expected to first decrease and then increase during succession. In fact, there is no contradiction between Vitousek and Reiners's hypothesis and the predictions of my model, for three main reasons.

The first reason is that Vitousek and Reiners's reasoning about nutrient inputs and outputs may be more relevant to ecosystem processes occurring on very long time scales that my model does not consider. My model shows that plant biomass can in principle build up while keeping nutrient inputs and outputs approximately in balance provided this balance is achieved fast enough that a quasiequilibrium is reached at any step of the species replacement sequence. This assumption is admittedly a simplification of reality that ignores the transient dynamics involved in the buildup of plant biomass, but the effect of this transient dynamics may be quite limited. Soil formation is another process of ecosystem development that occurs on longer time scales and that is much more likely to generate long-term imbalances between nutrient inputs and outputs. Soils can store large amounts of nutrients for long periods of time in forms that are unavailable to plants, especially in temperate and boreal regions. Although the dynamics of nutrient cycling in soils is complex (Agren and Bosatta 1996), nutrient storage in soils may be viewed as a sink of available nutrient on the shorter time scales of plant–decomposer interactions, hence it is treated implicitly as a loss of nutrient in organic form in my model. Thus, long-term imbalance between nutrient inputs and outputs due to storage of nutrients in unavailable form does not contradict short-term balance and increased cycling efficiency in available form.

This leads me to consider a second reason why Vitousek and Reiners's hypothesis does not contradict the predictions of my model. Vitousek and

Reiners implicitly considered that nutrients are lost from terrestrial ecosystems in dissolved inorganic form. In reality, there are many other sources of nutrient losses in most ecosystems. Nutrients can be lost in the form of dissolved organic matter, as in many forest ecosystems (Perakis and Hedin 2002), through fires, as in the savanna ecosystems discussed earlier in this chapter, or through denitrification and other gas losses. Therefore, there is no contradiction between the fact that mature ecosystems can have minimal losses of plant-available inorganic nutrients (Hedin et al. 1995), as predicted by my model, and the fact that total nutrient outputs can sometimes increase to balance nutrient inputs during ecosystem development.

There is a lesson of broader significance in this conclusion. It is common practice in ecology to interpret ecological models literally. Much confusion, however, can arise from a literal interpretation of compartmental ecosystem models in which nutrients undergo qualitative transformations between inorganic, dead organic, and living organic forms. For example, one would be tempted to infer from model (6.7) that the only source of nutrient loss in inorganic form is the output from compartment N. This interpretation, however, would be incorrect. Although compartment N is the only model compartment that is explicitly inorganic, it truly represents only that fraction of the inorganic nutrient that is available to plants since, by construction of the model, it is consumed and controlled by plants. Therefore, any other source of nutrient loss in inorganic form is accounted for by losses from other compartments. For instance, some nutrient can be lost in organic or inorganic form as the result of fire burning plant biomass; this should be accounted for as losses from the plant compartment. Similarly, some inorganic nutrient can be lost via leaching or denitrification during the decomposition process before it is made available to plants; this should be accounted for as losses from the decomposer compartment. Distinguishing between the form in which a nutrient is lost and the compartment from which it is lost is essential to make relevant interpretations of model predictions. Bob Holt and I showed concretely how the constraint of top-down control of inorganic nutrient by plants and the mass-balance constraint on inputs and outputs of inorganic nutrient can easily conflict and lead to implausible predictions of indefinite plant growth if other nutrient losses are ignored (Loreau and Holt 2004).

There is a third reason why Vitousek and Reiners's hypothesis does not contradict the predictions of my model. In agreement with E. P. Odum (1969), my model predicts that the ecosystem's nutrient cycling efficiency should increase during succession. In contrast, Vitousek and Reiners predicted that nutrient outputs should increase relative to nutrient inputs

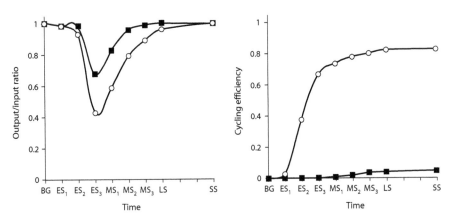

FIGURE 6.12. Changes in the ratio between total nutrient outputs and total
nutrient inputs (left) and in nutrient cycling efficiency (right) predicted by a
simple two-compartment model during a hypothetical successional sequence.
The horizontal axis represents time with several typical seral stages. Open circles: ecosystem with significant nutrient cycling. Filled squares: ecosystem
with virtually no nutrient cycling. Modified from Finn (1982).

during succession. But these two predictions address fundamentally different questions. Finn (1982) clarified the relationships between various indi-
cators of nutrient cycling and showed, using a simple two-compartment
model, that the ratio between total nutrient outputs and total nutrient inputs is expected to reach a minimum before increasing again irrespective of
the intensity of nutrient cycling, while nutrient cycling efficiency is expected
to gradually increase up to a level that reflects the intensity of nutrient cycling within the ecosystem during a successional sequence (figure 6.12).
Thus, there is no necessary relationship between these two measures of nutrient dynamics.

Ecosystem development also involves long-term changes in nutrient inputs, which are not considered in any of the theories and hypotheses discussed so far. Nutrient inputs are partly controlled by the biota through
such processes as rock weathering and nitrogen fixation. Because nitrogen
is absent from most primary substrates, nitrogen is often the limiting nutrient early in soil development in terrestrial ecosystems. Biological fixation
of atmospheric nitrogen then plays an important role. But this role decreases as succession proceeds and nitrogen is accumulated in the system,
leading often to replacement of nitrogen-fixing plants by nonfixing species
that are more efficient at using soil nitrogen. Despite these well-documented
biotic changes, total inputs of nitrogen stay relatively constant throughout
the development of terrestrial ecosystems (Vitousek 2004).

By contrast, inputs of phosphorus change markedly on longer time scales that range from millennia to millions of years. Because phosphorus is present in rocks and lacks a significant gas phase, it is made available through rock weathering and accumulates very much like nitrogen in the early development of terrestrial ecosystems. But once it has been released from primary substrates, its inputs decline (Vitousek 2004). Its availability then depends largely on internal recycling within the system. Although recycling of phosphorus is often very efficient, some leakages are inevitable, and over long periods of time these small leakages accumulate to generate significant phosphorus losses. The long-term decline in phosphorus inputs is thought to make terrestrial ecosystems switch from nitrogen limitation to phosphorus limitation as they age, ultimately resulting in loss of productivity (Walker and Syers 1976; Wardle et al. 2004). Thus, patterns of ecosystem development on very long time scales are likely governed by changes in nutrient inputs driven by abiotic factors.

CONCLUSION

In this chapter, I have shown that material cycling is a key ecosystem process that binds all ecosystem components together and transmits positive indirect effects to all of them. At equilibrium, primary and secondary productions obey simple, general laws. Equilibrium nutrient fluxes are determined by the parameters that govern the input–output balance of the limiting nutrient at the scale of the whole ecosystem. In particular, equilibrium primary production is equal to the net supply of inorganic nutrient to plants times the average number of times the limiting nutrient is cycled by the living compartments and their dependent detritus. Secondary production, which is derived from primary production, obeys similar laws. An important consequence of material cycling is that the productions of the various living compartments are coupled, and this tends to generate indirect mutualism between ecosystem components. Whether this indirect mutualism is strong enough to outweigh negative direct trophic effects between consumers and their resources depends both on the characteristics of the organisms involved and on those of the ecosystem as a whole.

Material cycling also mediates the effects of horizontal diversity within a trophic level on the properties of other trophic levels and plays a key role in ecosystem development. The combined dynamics of nutrient cycling and interspecific competition is able to explain general trends in ecosystem properties during succession, including increased biomass, increased production, decreased productivity per unit biomass, and increased nutrient cycling efficiency. Patterns on longer time scales, from millennia to millions of years, however, are largely governed by changes in nutrient inputs driven by abiotic factors.

The evolutionary consequences of the indirect ecological effects mediated by material cycling are a fascinating topic, but one that is inherently complex because of the very nature of the evolutionary process. Therefore, I defer their study to chapter 8 within the broader context of the evolution of ecosystems and ecosystem properties. I shall then revisit the plant–decomposer and plant–herbivore interactions in that context.

Spatial Dynamics of Biodiversity and Ecosystem Functioning:

Metacommunities and Metaecosystems

A defining feature of ecology over the last few decades has been a growing appreciation of the importance of considering processes operating at spatial scales larger than that of a single locality, from the scale of the landscape to that of the region (Ricklefs and Schluter 1993; Turner et al. 2001). Spatial ecology, however, has reproduced the traditional divide within ecology between the perspectives of population and community ecology on the one hand and ecosystem ecology on the other hand.

The population and community ecological perspective has focused on population persistence and species coexistence in spatially distributed systems (Hanski and Gilpin 1997; Tilman and Kareiva 1997) and has a strong background in theoretical ecology and simple, generic mathematical models. The metapopulation concept has occupied a prominent role in the development of this perspective (Hanski and Gilpin 1997). A *metapopulation* is a regional set of local populations that are spatially distinct but connected by dispersal, such that each population undergoes a dynamics of local extinction and colonization from elsewhere. The strength of this concept has been its ability to deliver specific testable hypotheses on the increasingly critical issue of conservation of fragmented populations in human-dominated landscapes. Because local extinction and colonization can be influenced by interspecific interactions such as predation and competition, a natural extension of the metapopulation concept is provided by the *metacommunity* concept, which designates a set of communities that are spatially distinct but connected by dispersal (Leibold et al. 2004). Significant new insights are being gained from the metacommunity perspective (Holyoak et al. 2005).

Another perspective, however, has developed from ecosystem ecology and is represented by landscape ecology. Landscape ecology is concerned with ecological patterns and processes in mosaics of nearby heterogeneous ecosystems (Turner 1989; Forman 1995; Pickett and Cadenasso 1995; Turner et al. 2001). It has a strong descriptive basis and a focus on whole-system properties, including abiotic processes. Models that address population persistence and conservation from this perspective are usually more detailed; they consider landscape structure and heterogeneity explicitly and therefore aim to be more realistic and directly applicable to concrete problems than the more general, abstract models of classical metapopulation and community ecology (e.g., Gustafson and Gardner 1996; With 1997).

Thus, the need to integrate the perspectives of community and ecosystem ecology is also apparent within the field of spatial ecology. The metacommunity concept has so far had an exclusive focus on the biotic components of ecosystems. Many critical issues at large spatial scales, however, require consideration of abiotic constraints and feedbacks to biotic processes. Nicolas Mouquet, Bob Holt, and I have recently proposed the metaecosystem concept as a theoretical framework for achieving this integration of the community and ecosystem perspectives within spatial ecology (Loreau et al. 2003b). A *metaecosystem* is defined as a set of ecosystems connected by spatial flows of energy, materials, and organisms across ecosystem boundaries. While the metacommunity concept considers connections among systems via the dispersal of organisms, the metaecosystem concept embraces all kinds of spatial flows among systems, including movements of inorganic nutrients, detritus, and living organisms, which are ubiquitous in natural systems. There has been considerable attention to impacts of spatial subsidies on local ecosystems (Polis et al. 1997). Such studies, however, are limited, in that a subsidy entering one local ecosystem must necessarily be drawn from another ecosystem and as such should have impacts on both source and target ecosystems. The properties of the higher-level system that arise from movements among coupled ecosystems have seldom been considered explicitly (Nakano and Murakami 2001).

In this chapter I revisit some of the themes of previous chapters in a spatial context that extends beyond the boundaries of a single community. A number of recent contributions have addressed the origin of spatial structure at various scales (Levin 1999; Solé and Bascompte 2006). Here I regard spatial structure at the regional scale as given, and I focus instead on some of the emergent properties that arise from spatial coupling of local communities and ecosystems and on the principles that govern them. I first show how metacommunity dynamics strongly links and constrains local diversity and regional diversity. I then consider the relationships between species diversity and ecosystem properties that emerge at local and regional scales from source–sink processes in metacommunities. Last, I explore the global constraints that spatial flows of materials across ecosystem boundaries impose on metaecosystem functioning.

LOCAL AND REGIONAL DIVERSITY

Historically, there have been two approaches to species diversity in community ecology: a local approach, based on niche theory, and a regional approach, based on the theory of island biogeography (MacArthur and Wilson 1967). In the local approach, interactions between competing species constrain local diversity, and coexistence is viewed as a result of differences in species' niches or life-history traits (chapter 2). In the regional approach, local dynamics is traditionally ignored, and local diversity is viewed as the result of the regional processes of immigration and extinction. In this theory, there are no limits to diversity except those arising from the size of the regional species pool (continent size) and the constraints on immigration events (continent–island distance). This apparent contradiction was called "MacArthur's paradox" (Schoener 1983b; Loreau and Mouquet 1999) because MacArthur's contributions were central in the development of both niche and island biogeography theories.

In reality the dynamics of species diversity at local and regional scales are not independent of one another. Theories of coexistence in spatially structured environments have started to break the borders between the local and regional approaches (chapter 2). More fundamentally, local diversity—also called α diversity—and regional diversity—also called γ diversity—are mutually dependent through β diversity, i.e., the diversity between communities. It is therefore impossible to understand local diversity, regional diversity, and the relationship between them without consideration of the dynamics that occurs across the two scales. Although this mutual dependency of local and regional diversity has been recognized in principle, it has generally been ignored in the interpretation of local-regional diversity relationships (Loreau 2000a). Even within metacommunity theory (Leibold et al. 2004), the species-sorting perspective (Leibold 1998) and much of neutral theory (Hubbell 2001) are based on the implicit assumption that local diversity is influenced by regional diversity, but there is no feedback of local diversity on regional diversity, just as in the classical theory of island biogeography. In a true metacommunity perspective, local and regional diversity should be emergent properties that arise from the dynamics of species interactions across scales and constrain each other. The source–sink metacommunity perspective Nicolas Mouquet and I have recently developed for spatially distributed competitive communities (Loreau and Mouquet 1999; Mouquet and Loreau 2002, 2003) shows precisely this.

To begin with, let us examine how regional horizontal diversity affects a given local community. Regional diversity influences the composition and diversity of the local community in at least two different ways: through a sampling effect and through a mass effect. The sampling effect was discussed within the context of biodiversity and ecosystem functioning in chapter 3. As the number of species present in a region increases, there is a higher probability of having the best-adapted species or combination of species in a local community. But there is also a higher probability of having species that fill specialized food, temporal, or spatial niches in the community, or that are more similar ecologically, thus fostering either stable or neutral coexistence. As a result, local species diversity tends to increase with regional species diversity.

Some species that colonize the community, however, may not be able to persist there indefinitely because of unfavorable environmental conditions or competitive exclusion by better-adapted species. Nevertheless, recurrent immigration from neighboring communities can maintain them locally if immigration is high enough to compensate for local population declines. Populations that are maintained by immigration despite negative local population growth rates are known as *sink* populations (Pulliam 1988), and the effect of sink populations on local species diversity is known as the *mass effect* (Shmida and Wilson 1985). In his classical microcosm experiments on predator–prey interactions, Gause (1934) had already demonstrated that immigration from an external source was able to maintain populations that otherwise would go extinct. In nature, a significant number of rare species are present in local communities because of immigrants coming from neighboring habitats, whether in plants (Shmida and Wilson 1985) or insects (Loreau 1994). Yet, despite long available empirical evidence, the mass effect has been analyzed theoretically only recently.

Nicolas Mouquet and I formalized it using a simple competitive lottery model with the addition of immigration from outside (Loreau and Mouquet 1999). Suppose a community of sessile organisms such as plants in which space occupancy obeys a competitive lottery in a homogeneous environment, as described by model (2.17) in chapter 2. Add an external source of immigrants in the form of a "propagule rain" as in island–continent models (Gotelli 1991; Gotelli and Kelly 1993), and our model then reads

$$\frac{dp_i}{dt} = (\alpha I_i + c_i p_i)V - m_i p_i, \tag{7.1a}$$

where

$$V = 1 - \sum_{j=1}^{S} p_j. \tag{7.1b}$$

Here, p_i is the proportion of sites occupied by species i. There are S such species that compete for a limited proportion of vacant sites, V. Each species i has a local potential recruitment rate c_i, a local mortality rate m_i, and a species-specific immigration rate I_i, which is determined by its long-distance dispersal capacity and its relative abundance in the regional source. Another parameter, α, encapsulates the overall immigration intensity into the community, which depends on the size of, and distance from, the regional source.

When the community is closed ($\alpha = 0$), a single species persists at equilibrium, the one with the highest basic reproductive capacity (chapter 2). All other species are driven to extinction by competitive exclusion. But as soon as the community is open to immigration ($\alpha > 0$), extinction no longer occurs for any of the species that receive external immigrants ($I_i > 0$) because their local populations are continuously replenished with these immigrants. Indeed, when their population is very small ($p_i \approx 0$), invasibility is guaranteed ($dp_i/dt > 0$) in equation (7.1). Stable coexistence then ensues (Loreau and Mouquet 1999). Although the stable coexistence of an arbitrary number of species sustained by immigration is possible in this deterministic model, many species will be lost because they are maintained at unrealistically low densities. If extinction due to demographic stochasticity or the finite number of sites in the community is taken into account by setting an extinction threshold for the proportion of sites occupied by each species, our model predicts that species richness at equilibrium increases continuously as immigration intensity, α, increases (figure 7.1, top). This increase, however, is not linear. Species richness on average starts increasing as soon as immigration intensity allows the potential recruitment of a number of individuals equivalent to the extinction threshold (here, a single individual in the community: figure 7.1, bottom). It then increases most steeply when immigration intensity is intermediate and reaches the ceiling set by regional richness when immigration is unrealistically high, i.e., when immigration contributes many more individuals than does local reproduction (figure 7.1).

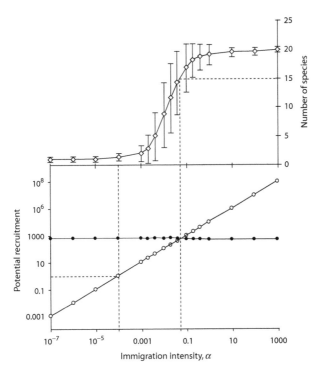

FIGURE 7.1. Species richness at equilibrium (mean ± 1 SD) vs. immigration intensity, α (top), and respective contributions of local reproduction (filled circles) and immigration (open circles) to the total potential recruitment in the community (bottom). The left dashed vertical line shows the immigration intensity allowing the potential recruitment of one immigrating individual on average in the community. The right dashed line shows the immigration intensity at which the local and regional contributions to total potential recruitment are equal. The potential species pool comprises 20 species. Modified from Loreau and Mouquet (1999).

Our model shows that immigration from an external source is able to maintain high local diversity in a system that would otherwise tend to competitive exclusion of all but one species. But the operation of this process requires that diversity be maintained in the source itself, i.e., at the regional scale. How is regional diversity maintained? This question can be answered only by looking at the metacommunity as a whole, in which each community acts both as a potential sink for immigrants dispersing from other communities and as a potential source of emigrants dispersing to other communities in the region.

The source–sink metacommunity model we built to address this question is a straightforward extension of the above model for multiple communities

(Mouquet and Loreau 2002, 2003). It incorporates spatial structure at two levels: within communities and between communities. At the local scale (within communities), it considers a competitive lottery in a set of identical sites, each of which can be occupied by a single individual, just as in the previous model. At the regional scale (between communities), dispersal is assumed to occur through a passive immigration–emigration process. Heterogeneity of environmental conditions at the regional scale is obtained by changing species-specific parameters in each community. This assumes that species exhibit different phenotypic responses in different communities as a result of different local environmental factors.

Define p_{ik} as the proportion of sites occupied by species i in community k. Here S species compete for a limited proportion of vacant sites, V_k, in each community k, and there are N such communities. Each species i is characterized by a set of reproduction–dispersal parameters b_{ilk}, which describe the rate at which new individuals are produced in community l and establish in community k. When $k = l$, b_{ilk} corresponds to local reproduction, and when $k \neq l$, b_{ilk} corresponds to dispersal from community l to community k. Each species i dies in community k at a mortality rate m_{ik}. When a species immigrates into a particular community, it takes the parameters corresponding to that community. The model then reads as follows:

$$\frac{dp_{ik}}{dt} = V_k \sum_{l=1}^{N} b_{ilk} p_{il} - m_{ik} p_{ik}, \tag{7.2a}$$

where

$$V_k = 1 - \sum_{j=1}^{S} p_{jk}. \tag{7.2b}$$

A necessary condition for there to be an equilibrium in this model is $S \leq N$ (Mouquet and Loreau 2002). Thus, there cannot be more species than communities in the metacommunity at equilibrium. This rule provides an equivalent to the competitive exclusion principle in a local community (chapter 2). At equilibrium each species must further satisfy the condition (Mouquet and Loreau 2002)

$$\overline{R}_i = \frac{\sum_{k=1}^{N} R_{ik} w_{ik}}{\sum_{k=1}^{N} w_{ik}} = 1, \tag{7.3}$$

where

$$R_{ik} = V_k^* r_{ik}, \tag{7.4}$$

$$r_{ik} = \frac{\sum\limits_{l=1}^{N} b_{ikl}}{m_{ik}}, \tag{7.5}$$

$$w_{ik} = \sum\limits_{l=1}^{N} b_{ilk} p_{il}^*. \tag{7.6}$$

Parameter r_{ik} can be interpreted as the local basic reproductive capacity of species i in community k [equation (7.5)]. Multiplying r_{ik} by the proportion of vacant sites at equilibrium, V_k^*, we obtain the local net reproductive capacity of species i in community k at equilibrium, R_{ik} [equation (7.4)]. Finally, w_{ik} is the total quantity of propagules produced by species i that arrive in community k per unit time at equilibrium [equation (7.6)]. Consequently, $\overline{R_{ik}}$ is the regional average net reproductive capacity of species i, weighted by the total quantity of propagules arriving in each community, at equilibrium [equation (7.3)]. Clearly, for the metacommunity to reach equilibrium, $\overline{R_{ik}}$ must be equal to 1 [equation (7.3)]; i.e., each individual of each species must produce one individual on average during its lifetime in the metacommunity as a whole.

Because all the regional average net reproductive capacities must be equal at equilibrium, this sets a constraint of regional similarity between coexisting species. The basic reproductive capacities of all species must be sufficiently balanced over the region for equality (7.3) to be possible. Local coexistence is then possible in a metacommunity when species are locally different but regionally similar with respect to their reproductive capacities. Local coexistence is explained by compensations among species' competitive abilities at the scale of the region. As a corollary, the net reproductive capacity, and hence also the basic reproductive capacity, of any species cannot be lower than that of any other species in all communities simultaneously. This condition requires habitat differentiation among species; i.e., the environment should be heterogeneous enough at the regional scale such that each species is competitively dominant in at least one community. Thus, in a source–sink metacommunity, the number of species that coexist locally and regionally is highest when species have different niches (*habitat differentiation constraint*), but similar competitive abilities (*regional similarity constraint*), at the scale of the region. These two constraints that arise at the regional scale parallel those found in classical competition theory in an isolated local community: habitat differentiation acts as a stabilizing mechanism, while regional similarity acts as an equalizing mechanism. Both types of mechanisms are necessary to allow species coexistence (chapter 2).

These rules place strong constraints on both local and regional species diversity. Within these constraints, however, a wide variation of local and regional diversity is possible, and this variation is driven in particular by changes in dispersal among communities. To highlight the effect of dispersal on species diversity, assume for simplicity that a proportion of the total reproductive output remains resident, while the rest emigrates through a regional pool of dispersers that are equally redistributed in all other communities, and that the proportions of dispersers (a) and nondispersers ($1 - a$) are equal for all species and all communities. Parameter a may thus also be interpreted as a measure of the relative importance of regional versus local dynamics. With these assumptions,

$$b_{ilk} = \begin{cases} (1 - a)c_{il} & \text{if } k = l, \\ \frac{a}{N-1}c_{il} & \text{if } k \neq l, \end{cases} \tag{7.7}$$

in equation (7.1). Here c_{il} is the potential reproductive rate of species i in community l, which encapsulates local reproduction, short-distance dispersal, and establishment capacities.

This model was simulated until equilibrium for a metacommunity consisting of 20 species competing in 20 communities, with a matrix of species' local basic reproductive capacities corresponding to a deviation of 5 percent from strict regional similarity. In the case of strict regional similarity the matrix is completely symmetrical with each species being the best competitor in one community. The three components of species diversity (α, β, and γ diversity) were related through the additive partition advocated by Lande (1996) and Loreau (2000a):

$$\gamma = \beta + \bar{\alpha}, \tag{7.8}$$

where $\bar{\alpha}$ is the mean α diversity of local communities. Although this additive partition differs from the classical, multiplicative approach (Whittaker 1972), it provides a unified framework for the measurement of diversity that is internally consistent and that can be applied to a nested hierarchy of multiple spatial scales.

Varying dispersal has a dramatic effect on the three components of diversity (figure 7.2). When dispersal is zero, local (α) diversity is minimum (one species), whereas between-community (β) diversity and regional (γ) diversity are maximum; in each community a different species is locally the best competitor. As dispersal increases to an intermediate value a_{max}, an increasing number of species are maintained by immigration above the extinction threshold, so that α diversity increases, while at the same time communities become more similar in composition, so that β diversity decreases.

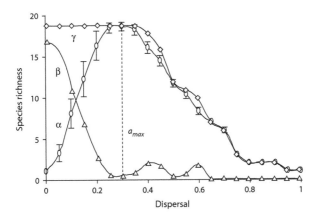

FIGURE 7.2. Local (α) diversity (circles), between-community (β) diversity (triangles), and regional (γ) diversity (diamonds) (means $\pm$ 1 SD across communities) as functions of dispersal (proportion of dispersers, a) in a source–sink competitive metacommunity. a_{max} is the dispersal value at which local species diversity is maximum. Modified from Mouquet and Loreau (2003).

Regional diversity, however, remains relatively constant. As dispersal increases above a_{max}, both local and regional diversity decrease, while β diversity stays close to zero because the best competitor at the scale of the region tends to dominate each community, and other species are progressively excluded. At high dispersal, the metacommunity functions effectively as a single community in which one species outcompetes all others. A hump-shaped relationship between local diversity and dispersal emerges from these constraints. In the ascending part of the curve, γ diversity is determined by regional environmental heterogeneity, and dispersal acts to transfer its effect from the between-community (β) to the within-community (α) component of diversity. In the descending part of the curve, dispersal leads to homogenization of the metacommunity, which has a negative effect on regional, and hence also local, diversity.

These predictions hold for metacommunities governed by source–sink dynamics. Interestingly enough, predictions are different for neutral metacommunities, i.e., metacommunities in which all species are competitively equivalent. Neutral metacommunities do not show the collapse of α and γ diversity at high dispersal rates (figure 7.3) because global competitive exclusion does not occur. Therefore, α diversity can only increase as dispersal redistributes species across the landscape. These contrasting predictions provide a simple potential test of the mechanisms that maintain species coexistence at local and regional scales. In a meta-analysis of 50 experiments

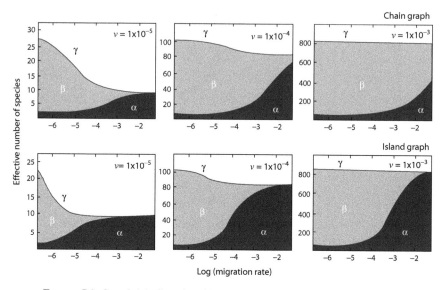

Figure 7.3. Local (α) diversity (black), between-community (β) diversity (gray), and regional (γ) diversity (black line) as functions of migration rate in a neutral competitive metacommunity for three values of the speciation rate (ν) and two spatial configurations (chain graph = linear chain of communities; island graph = network where every community is connected to every other community). Modified from Economo and Keitt (2008).

that manipulated dispersal in plant and animal communities, Cadotte (2006) found support for some of the theoretical predictions of metacommunity theory. In particular, he showed that dispersal consistently increases local species diversity, as predicted by both source–sink and neutral metacommunity theories. He also showed that a hump-shaped statistical model best fitted available data from animal experiments on the relationship between local diversity and dispersal, thus supporting the prediction of source–sink metacommunity theory but contradicting that of neutral metacommunity theory.

The above results show that dispersal is a major determinant of the relationship between local and regional diversity. To further explore this issue, we varied maximum regional species richness by varying the degree of regional environmental heterogeneity in the metacommunity for each dispersal value (Mouquet and Loreau 2003). Variation in environmental heterogeneity was obtained by defining a parameter E_k measuring the environmental condition of community k in a range from 0 to 1, and a parameter H_i measuring the niche preference of species i to environmental conditions, also in a range from 0 to 1. The potential reproductive rate of species i

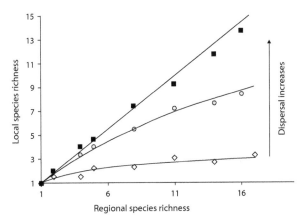

FIGURE 7.4. Relationships between local and regional species richness for various values of dispersal in a competitive metacommunity ($a = 0.1$, black squares; $a = 0.075$, gray circles; $a = 0.025$, white diamonds). For each dispersal value the gradient of regional species richness was obtained by varying the degree of regional heterogeneity. These results were obtained for low to intermediate dispersal values. At high dispersal, local and regional diversities are equal and the relationship is linear. Modified from Mouquet and Loreau (2003).

in community k, c_{ik}, was assumed to be greater as its niche optimum was closer to the local environmental condition ($c_{ik} = (1 - |E_k - H_i|) \times 3$). Variation of regional environmental heterogeneity was then generated by varying the distribution of E_k values across communities. Figure 7.4 shows the resulting relationships between local and regional diversity for various levels of dispersal. When dispersal is low, local species richness is limited by the locally dominant species irrespective of regional species richness, and the resulting relationship between local and regional diversity is saturating. When dispersal is higher, local species richness becomes equal to regional species richness, and the relationship between local and regional diversity is linear.

Local vs. regional diversity plots have often been interpreted as indicative of community saturation; unsaturating linear curves would be typical for "unsaturated," "noninteractive" communities, whereas saturating curves would indicate "saturated," "interactive" communities (Terborgh and Faaborg 1980; Cornell and Lawton 1992; Cornell 1993). The above results show that this interpretation is unwarranted since competition for space was equally strong in all cases. The shape of local-regional richness relationships does not tell us anything about the strength of species interactions or community saturation arising from these interactions. In fact, the very concept of community saturation is ambiguous as it may variously denote saturation of species richness through time for a given regional species pool, saturation of species richness with respect to variations in species pools across regions, or saturation of total density, biomass, or other communitywide properties, and species interactions affect these various forms of saturation differently (Loreau 2000a). The shape of local-regional richness relationships is fundamentally related to the scale at which a local community is defined and the dispersal properties of the organisms considered. Generally speaking, expanding the scale at which local communities are defined amounts to transferring the environmental heterogeneity that is responsible for the bulk of diversity from the regional to the local scale, hence from β to α diversity. Increasing dispersal across the landscape has a similar effect. The source–sink metacommunity perspective that we developed allows studying these scale and dispersal dependencies quantitatively.

The theory reviewed here is focused on the maintenance of horizontal diversity in source–sink metacommunities. There is growing recognition that the assembly and dynamics of food webs should also be understood in a spatial context. Although recent work offers promising avenues to tackle this issue (Holt 2002; McCann et al. 2005), we are still far from a full-fledged theory of species diversity in food-web metacommunities. Extending metacommunity theory to systems with multiple trophic levels is an exciting challenge for future research.

SPECIES DIVERSITY AND ECOSYSTEM
PROPERTIES IN METACOMMUNITIES

We have seen in chapter 3 that the traditional approach to diversity–productivity relationships has been to regress species diversity on productivity—or, more exactly, on factors, such as climate and soil fertility, that determine productivity—across sites with different environmental characteristics. In contrast, recent experimental and theoretical work has focused on the specific effect of species diversity on productivity when all other factors are held constant. The two approaches have led to different results, which can be reconciled by recognizing that they address different causal relationships at different scales.

But diversity–productivity relationships are also expected to depend on the "kind of diversity" present in a community, i.e., on the coexistence mechanisms that are responsible for the maintenance of diversity within the community (Mouquet and Loreau 2002). Different coexistence mechanisms involve different environmental and evolutionary constraints on organisms, and these constraints shape both the diversity and the productivity of the communities and ecosystems these organisms form. The relationships between diversity and ecosystem properties such as productivity then emerge as products of environmental and evolutionary constraints, in which diversity determines ecosystem properties as much as ecosystem properties determine diversity. What relationships between species diversity and ecosystem properties emerge from source–sink processes in metacommunities?

Before addressing this relatively complex question, let us again examine first how immigration from an external source affects ecosystem properties in a local community. This is a simpler issue, which allows a few basic principles to be uncovered. Take model (7.1) above, which represents a competitive lottery for vacant sites subject to a propagule rain. As immigration intensity, α, increases, the proportion of vacant sites decreases, leading to better space occupation and increased plant cover because the propagule rain increases the amount of new individuals available to occupy sites made vacant by mortality (Loreau and Mouquet 1999). Despite improved space occupation, however, community productivity, or primary production, often decreases or at best increases very slightly as immigration intensity increases, depending on the assumed relationship between a species' specific immigration rate, I_i, and its potential reproductive rate, c_i (figure 7.5). Here, each species' productivity is approximated by its local recruitment potential as measured by its potential recruitment rate at a site times the proportion of sites it occupies. Thus, total plant production (primary production) is approximated by

$$\Phi = \sum_{i=1}^{S} c_i p_i. \tag{7.9}$$

Three scenarios are considered in figure 7.5: species-specific parameters I_i and c_i are negatively correlated (figure 7.5A), independent (figure 7.5B), or positively correlated (figure 7.5C).

Why does primary production often decline, and at best increase only slightly, as immigration intensity increases? To understand this, it is important to realize that the mass effect as defined by Shmida and Wilson (1985) is really a combination of two effects that have contradictory functional consequences. First is the *mass effect per se*, i.e., the flow of immigrants into the community. Second, there is a *dilution effect*; i.e., the local best competitor is diluted into a mass of less adapted species that immigrate from more favorable environments elsewhere.

The mass effect per se contributes to increase space occupation by the community, and hence also its productivity, because it increases the

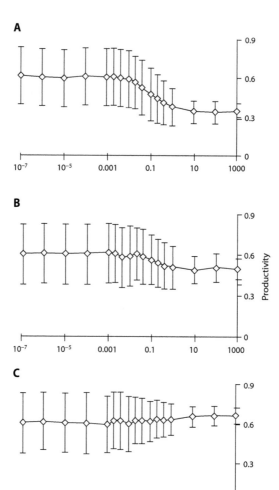

FIGURE 7.5. Primary productivity (mean ± 1 SD) vs. immigration intensity, α, under three scenarios for the relation between the species-specific immigration rate, I, and the potential reproduction rate, c: (A) I is a decreasing function of c ($I = 1 - c$); (B) I and c are independent; (C) I is an increasing function of c ($I = c$). Modified from Loreau and Mouquet (1999).

recruitment of new individuals that can occupy vacant sites. To separate it from the dilution effect, assume that all immigrants belong to the locally best competitor, i.e., to the species with the highest basic reproductive capacity. All other species are then driven to extinction, just as in a closed community. Increasing immigration intensity increases space occupation by the best competitor without affecting its local potential recruitment rate; hence it also increases primary production [equation (7.9)]. Thus, the mass effect per se has a positive effect on all aggregate community properties.

But the mass effect is accompanied by immigration of other species, which are less adapted to local conditions and persist through immigration despite their local competitive disadvantage. These species contribute to decrease the proportion of sites occupied by the best competitor and hence to decrease both space occupation and primary production compared with the community in which only the best competitor experiences immigration. This dilution effect partly counteracts the mass effect per se, so that their combined effect on productivity is negative or only slightly positive (figure 7.6).

These conclusions on the functional consequences of immigration in communities of sessile organisms competing for space can be generalized to multitrophic ecosystems in which horizontal diversity is maintained by immigration. Nutrient inputs or immigration at any trophic level contribute to increase the biomass and productivity of that trophic level, sometimes with counterintuitive effects on the biomass and productivity of other trophic levels (Loreau and Holt 2004). Thus, the mass effect per se is positive for the functioning of the trophic level where it occurs. But at the same time, the dilution effect that comes with the immigration of other species counteracts competitive exclusion and species sorting at that trophic level.

To see this, consider a simple food chain like that studied in chapter 4, in which one trophic level i is subject to immigration from an external source. Let S_i be the total number of species from trophic level i, B_{ij} the biomass of species j from trophic level i, g_{ij} its local mass-specific growth rate (which is a function of the biomasses of the other trophic levels in the ecosystem), I_{ij} its proportional share in the flow of immigrants at trophic level i, and α_i the overall immigration rate at trophic level i. The population dynamics of each species j from trophic level i subject to immigration is then described by the equation

$$\frac{dB_{ij}}{dt} = \alpha_i I_{ij} + g_{ij} B_{ij}. \qquad (7.10)$$

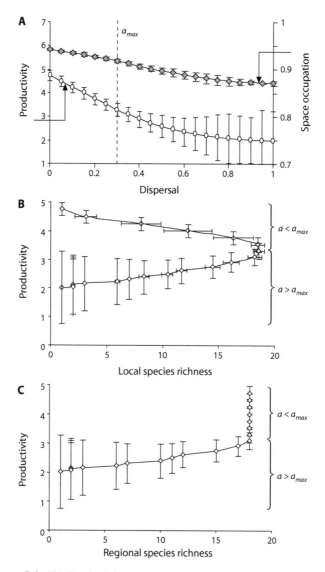

FIGURE 7.6. (A) Productivity (circles) and space occupation (diamonds) (mean ± 1 SD across communities) vs. dispersal (proportion of dispersers, a) in a competitive metacommunity. (B) The relationship between local species richness and local productivity when dispersal varies. (C) The relationship between regional species richness and local productivity when dispersal varies. a_{max} is the dispersal value at which species richness is maximum. Modified from Mouquet and Loreau (2003).

Summing over all species from that trophic level yields

$$\frac{dB_i}{dt} = \alpha_i + \overline{g}_i B_i, \tag{7.11}$$

where $B_i = \sum_{j=1}^{S_i} B_{ij}$ is the total biomass of trophic level i, and

$$\overline{g}_i = \frac{\sum\limits_{j=1}^{S_i} g_{ij} B_{ij}}{\sum\limits_{j=1}^{S_i} B_{ij}} \tag{7.12}$$

is the average local mass-specific growth rate of trophic level i.

If there are no trade-offs or niche differences that maintain local coexistence within this trophic level, the species with the highest local mass-specific growth rate for given biomasses of the other trophic levels will eventually outcompete all the other species in the absence of immigration, just as in the R* rule for a single trophic level (chapter 2). Assuming that species are ranked such that species 1 has the highest local mass-specific growth rate and species S_i has the lowest, we then see immediately that $g_{i1} > \overline{g}_i$ as soon as more than one species are present. Thus, the dilution effect counteracts species sorting because it maintains weaker competitors that would be displaced in its absence. Any ecosystem property that is enhanced by species sorting will be deteriorated by the dilution effect, while any ecosystem property that is deteriorated by species sorting will be enhanced by the dilution effect.

For example, consider the total biomass of the trophic level subject to immigration in model (7.10). At equilibrium, this biomass is [equation (7.11)]

$$B_i^* = \frac{\alpha_i}{-g_i^*}. \tag{7.13}$$

Note that the average local mass-specific growth rate must be negative at equilibrium to balance immigration. Since $g_{i1}^* > \overline{g}_i^*$, $-g_{i1}^* < -\overline{g}_i^*$, and the total biomass of trophic level i would be higher if only species 1 were present. Thus, the total biomass of a trophic level subject to immigration is increased by species sorting and decreased by the dilution effect, just as in a single-trophic-level competitive community. The functional consequences of the dilution effect on other trophic levels, however, varies depending on the configuration of the ecosystem and the specific traits that differ among species, just as with species sorting (Holt and Loreau 2001).

How do these conclusions extend to metacommunities, in which immigration in a community is generated by emigration from other communities? Let us first consider our metacommunity model described by equations (7.2) and (7.7), which is a straightforward extension of model (7.1) for a single community obeying a competitive lottery and subject to immigration. Using the same proxies as before for space occupation (proportion of sites occupied) and productivity [equation (7.9)], numerical simulations show that average productivity and space occupation across the metacommunity decrease as dispersal increases (figure 7.6A). This is explained easily by the principles I have brought out above. The mass effect per se plays a minor role at the scale of the whole metacommunity because on average local gains of individuals through immigration are compensated for by local losses through emigration. On the other hand, the dilution effect increases with dispersal because dispersal shuffles species around and hence increases the proportion of poorly adapted species in all communities. The preeminence of the dilution effect leads to deteriorated community properties across the metacommunity. When this result is combined with the hump-shaped relationship between local species diversity and dispersal (figure 7.2), a hump-shaped relationship also emerges between average productivity and local species richness (figure 7.6B). At the regional scale, however, the relationship between average productivity and regional species richness is either positive or absent (figure 7.6C) because regional species richness is constant or decreases with increasing dispersal (figure 7.2).

These results provide theoretical support to the hypothesis that different diversity–productivity relationships may emerge at different spatial scales, although the mechanisms involved here are different from those proposed in other studies (Bond and Chase 2002; Chase and Leibold 2002). Using a verbal model, Bond and Chase (2002) suggested that regional complementarity among species could lead to a positive relationship between productivity and regional species richness. In contrast, a hump-shaped relationship would be found at the local scale because local species richness would increase first through local niche complementarity (generating a positive relationship with productivity) and then through a source–sink effect (generating a negative relationship with productivity). Our results confirm their intuition, but they involve no local niche complementarity. Both the local hump-shaped and the regional positive diversity–productivity relationships arise from pure source–sink metacommunity processes in our work.

New patterns, however, emerge in fluctuating environments. Nicolas Mouquet, Andy Gonzalez, and I developed a more mechanistic consumer–resource model to explore the effects of species diversity on ecosystem productivity and its temporal stability in metacommunities under fluctuating environmental conditions (Loreau et al., 2003a; Gonzalez et al. 2009). This model makes similar assumptions as the previous one, in particular, the fact that dispersal is global and identical for all species, and dispersers are redistributed uniformly across the landscape. The main differences lie in the presence of an explicit consumer–resource local interaction, which allows a more straightforward measurement of productivity, and the presence of environmental fluctuations. The model reads as follows:

$$\frac{dN_{ij}(t)}{dt} = [e_{ij}c_{ij}(t)R_j(t) - m_{ij}]N_{ij}(t) + \frac{a}{M-1}\sum_{\substack{k=1 \\ k \neq j}}^{M} N_{ik}(t) - aN_{ij}(t),$$

$$\frac{dR_j(t)}{dt} = I_j - q_jR_j(t) - R_j(t)\sum_{i=1}^{S}c_{ij}(t)N_{ij}(t),$$

(7.14)

where $N_{ij}(t)$ is the biomass of species i (e.g., a plant), and $R_j(t)$ is the amount of limiting resource (e.g., a nutrient such as nitrogen) in community j at time t. The metacommunity consists of M communities and S species in total. Species i consumes the resource at a rate $c_{ij}(t)$, converts it to new biomass with efficiency e_{ij}, and dies at rate m_{ij} in community j. The resource is renewed locally through a constant input flux I_j and is lost at a rate q_j. All species disperse at a rate a. Consumption rates $c_{ij}(t)$ vary as local environmental conditions change through time and are assumed to reflect the matching between species traits and environmental conditions as in the previous section. Again defining H_i as the constant trait value of species i and $E_j(t)$ as the fluctuating environmental value of community j (both varying between 0 and 1), consumption rates are given specifically by

$$c_{ij}(t) = \frac{1.5 - |H_i - E_j(t)|}{10}.$$

(7.15)

Fluctuations of local environmental values are assumed to be sinusoidal with period T:

$$E_j(t) = \frac{1}{2}\left[\sin\left(E_j(0) + \frac{2\pi t}{T}\right) + 1\right].$$

(7.16)

Ecosystem productivity at time t is defined as the production of new biomass per unit time, which, averaged across the metacommunity, is

$$\Phi(t) = \frac{\sum_{i=1}^{S}\sum_{j=1}^{M}e_{ij}c_{ij}(t)R_j(t)N_{ij}(t)}{M}.$$

(7.17)

The results shown in figure 7.7 are for a metacommunity made up of 20 species and 20 communities, subject to three levels of resource input to examine the effects of ecosystem fertility. The initial environmental conditions $E_j(0)$ in each community were chosen at random from a uniform distribution between 0 and 1, and the period of environmental fluctuations was large enough so that there was rapid competitive exclusion in the absence of dispersal.

This model leads to the same hump-shaped relationship between local (α) diversity and dispersal as does the previous one (figure 7.7B). When dispersal is absent, each community is isolated, and the species best adapted to the initial local environmental conditions excludes all other species. Regional (γ) diversity is then high and is affected only by the rate of resource input (figure 7.7A). As dispersal increases, local diversity first increases and then decreases as the metacommunity is gradually homogenized. When dispersal is very high, the metacommunity reduces to a metapopulation of a single species, the one that is best adapted to the average environmental conditions at the regional scale. Regional diversity then drops correspondingly.

In contrast to the previous model, average productivity now follows a hump-shaped pattern similar to that of local species diversity (figure 7.7C), while the coefficient of variation of productivity follows an inverse pattern (figure 7.7D). The positions of the peak of average productivity and of the trough of variability in productivity are little affected by the rate of resource input, although average productivity increases significantly, and variability in productivity decreases, with increasing resource input. Thus, variations in dispersal rate generate strongly nonlinear, parallel variations in local species diversity, average productivity, and the stability (sensu reduced variability) of productivity. This generates a positive relationship between productivity and local diversity (figure 7.7E) and a negative relationship between variability in productivity and local diversity (figure 7.7F).

Differences from the previous model are due to two new mechanisms that arise from the combination of spatial and temporal environmental heterogeneity: (1) compensatory fluctuations between different species, phenotypes, or functional groups in the presence of spatiotemporal environmental heterogeneity, which generate spatial insurance effects of biodiversity; and (2) spatial averaging of spatiotemporal environmental heterogeneity.

Biodiversity acts as biological insurance for local ecosystem functioning by allowing functional compensations between species, phenotypes, or functional groups in time. These insurance effects include an increase in the temporal mean of productivity when there is selection for adaptive responses to environmental fluctuations, and a decrease in its temporal variability because of temporal complementarity between species environmental responses (chapter 5). Here, however, these effects occur despite the fact that local coexistence is impossible, and thus no temporal insurance effect can occur within a closed system. Therefore, insurance effects shown by this model are generated by the spatial dynamics of the metacommunity. When different systems experience different environmental conditions and fluctuate asynchronously, different species thrive in each system at each point in

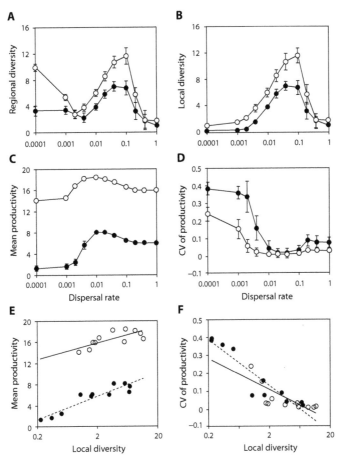

FIGURE 7.7. Mean regional (γ) diversity (A), mean local (α) diversity (B), mean ecosystem productivity (C), and temporal coefficient of variation of ecosystem productivity as functions of dispersal rate averaged over 50 simulations in a source–sink competitive metacommunity in fluctuating environments described by model (7.14)–(7.17). Mean ecosystem productivity (E) and temporal coefficient of variation of ecosystem productivity (F) as functions of alpha diversity. Means of 50 simulations are shown with standard error (A–C) or standard deviation (D). The symbols in each panel indicate different rates of resource input: filled circles, $I = 110$; empty circles, $I = 165$. Modified from Gonzalez et al. (2009).

time, and dispersal ensures that the species adapted to the new environmental conditions locally are available to replace less adapted ones as the environment changes. As a result, biodiversity enhances and buffers ecosystem processes by virtue of spatial exchanges among local systems in a heterogeneous landscape, even when such effects do not occur in a closed homogeneous system. This is the *spatial insurance* hypothesis (Loreau et al. 2003a).

As figure 7.7 shows, spatial insurance effects are strongly dependent on dispersal rate, which determines metacommunity connectivity. Local species diversity and the insurance effects it generates are highest at an intermediate dispersal rate and collapse at both low and high dispersal rates. At both ends of the dispersal gradient, functional compensations and adaptive shifts between species are prevented, leading to relatively low average productivity as well as large fluctuations in productivity as the single surviving species tracks environmental fluctuations.

Spatial averaging of spatiotemporal environmental heterogeneity is a direct result of the homogenizing effect of dispersal among communities, an effect that is independent of species diversity. Although average productivity decreases and the variability of productivity increases, as dispersal increases beyond its diversity-maximizing value, average productivity is higher and variability in productivity is lower, in a highly connected system than in a poorly connected system (figure 7.7C and D). Dispersal tends to enhance mean population density, and hence average productivity, through a spatial storage effect (Ives et al. 2004). Decreased variability in productivity occurs because the intermediate-type species that dominates a highly connected system is always able to find suitable conditions somewhere in the landscape and averages out environmental variations across the various local sites. Detailed analyses show that spatial averaging has a roughly constant effect beyond an intermediate threshold dispersal rate and plays a significant role mainly when local species diversity declines at high dispersal rates. In contrast, the spatial insurance effects of species diversity are highest at the intermediate dispersal rate that maximizes local diversity (Loreau et al. 2003a).

Several recent experiments have addressed the effects of dispersal on the relationship between biodiversity and ecosystem functioning (Gonzalez and Chaneton 2002; France and Duffy 2006; Matthiessen and Hillebrand 2006; Venail et al. 2008). The results of these experiments are broadly consistent with the spatial insurance hypothesis, although none represents a precise test of the model we used. For instance, Matthiessen and Hillebrand (2006) constructed laboratory metacommunities of benthic microalgae, and enhanced the rate of dispersal from the experimental "regional

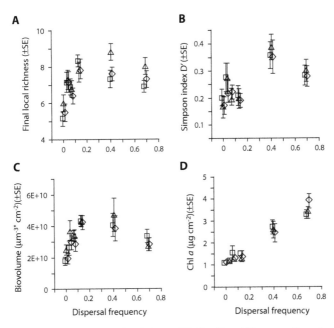

FIGURE 7.8. Final local species richness (A), final local Simpson diversity (B), local biovolume (C), and local chlorophyll *a* content (D) versus dispersal frequency in laboratory metacommunities of benthic microalgae. Biovolume and chlorophyll *a* content are two measures of algal biomass. Squares display the initial diversity of two species, triangles the initial diversity of four species, and diamonds the initial diversity of eight species. Modified from Matthiessen and Hillebrand (2006).

pool" (aquaria) into the local communities (open-top, upright plastic tubes in the aquaria) by increasing the frequency at which the algae were scraped from the bottom of the aquaria and resuspended in the water column. As predicted by the spatial insurance hypothesis, they found unimodal relationships between dispersal rate and both local species richness and biovolume (a measure of algal biomass) (figure 7.8), resulting in a positive relationship between species richness and biomass. However, no attempt was made to control spatiotemporal environmental heterogeneity and study community dynamics in this relatively short-term microcosm experiment. The fact that slightly different mechanisms yield qualitatively similar humpshaped relationships between dispersal and local species diversity and between dispersal and biomass or productivity in both models and experiments suggests that these relationships may be relatively robust metacommunity properties.

MATERIAL FLOWS AND ECOSYSTEM
FUNCTIONING IN METAECOSYSTEMS

Flows of nutrients, whether in the form of inorganic elements, detritus, or living organisms, can exert major influences on the functioning of local ecosystems (Polis et al. 1997; Loreau and Holt 2004; Leroux and Loreau 2008). Less appreciated is the fact that these flows may also impose global constraints at the scale of the metaecosystem as a whole, thereby generating a strong interdependence among local ecosystems.

To highlight these constraints, I concentrate on the simplest possible model of a closed, nutrient-limited metaecosystem consisting of two connected local ecosystems, 1 and 2, each of which in turn consists of two interacting compartments, plants (with nutrient stock P) and inorganic nutrients (with stock N). Spatial flows among ecosystems are assumed to occur among similar compartments (i.e., from inorganic nutrient to inorganic nutrient, and from plants to plants). They are also assumed to be independent of local interactions among ecosystem compartments, such that spatial flows and local growth rate are additive in the dynamical equation for each ecosystem compartment. Let F_{Xij} denote the directed spatial flow of nutrient stored in compartment X from ecosystem i to ecosystem j, Φ_i primary production in ecosystem i, and R_i the flow of recycled nutrient within ecosystem i. Local and global mass balance leads to the following set of equations describing the dynamics of the metaecosystem:

$$\frac{dN_1}{dt} = F_{N21} - F_{N12} - \Phi_1 + R_1, \tag{7.18a}$$

$$\frac{dN_2}{dt} = F_{N12} - F_{N21} - \Phi_2 + R_2, \tag{7.18b}$$

$$\frac{dP_1}{dt} = F_{P21} - F_{P12} + \Phi_1 - R_1, \tag{7.18c}$$

$$\frac{dP_2}{dt} = F_{P12} - F_{P21} + \Phi_2 - R_2. \tag{7.18d}$$

This description in terms of directed flows among ecosystems and compartments (figure 7.9A) can be reduced to a simpler description in terms of net flows as follows:

$$\frac{dN_1}{dt} = F_N - G_1, \tag{7.19a}$$

$$\frac{dN_2}{dt} = -F_N - G_2, \tag{7.19b}$$

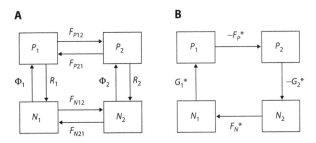

FIGURE 7.9. Material flows in a closed nutrient-limited metaecosystem consisting of two connected local ecosystems, 1 and 2, each of which in turn consists of two interacting compartments, plants (P) and inorganic nutrients (N): (A) directed nutrient flows, and (B) net nutrient flows at equilibrium. Modified from Loreau et al. (2005).

$$\frac{dP_1}{dt} = F_P + G_1, \tag{7.19c}$$

$$\frac{dP_2}{dt} = -F_P + G_2, \tag{7.19d}$$

where $F_X = F_{X21} - F_{X12}$ is the net spatial flow of nutrient of compartment X from ecosystem 2 to ecosystem 1, and G_i is net local plant growth in ecosystem i.

Note that both local mass conservation in the absence of spatial flows and global mass conservation are met, as required for closed systems. Additional constraints emerge from spatial coupling of local ecosystems as the metaecosystem reaches equilibrium. At equilibrium the left-hand side of equations (7.18) and (7.19) vanishes, which imposes

$$F_N^* = -F_P^* = G_1^* = -G_2^*, \tag{7.20}$$

where asterisks denote functions evaluated at equilibrium. This set of equalities can be interpreted as a double constraint: (1) a *source–sink constraint within ecosystem compartments*: for each compartment, positive growth in one ecosystem must be balanced by negative growth in the other ecosystem at equilibrium, which means that one ecosystem must be a source whereas the other must be a sink for that compartment; and (2) a *source–sink constraint between ecosystem compartments*: the total net spatial flow across the boundaries of each ecosystem must vanish at equilibrium, which means that one compartment must be a source whereas the other must be a sink.

These constraints can easily be generalized to closed metaecosystems with an arbitrary number of local ecosystems and an arbitrary number of ecosystem compartments (Loreau et al. 2003b). Global mass balance imposes that the local growth rates of each ecosystem compartment sum to zero across the various ecosystems that make up the metaecosystem, yielding a generalized source–sink constraint within ecosystem compartments. Similarly, global mass balance imposes that the net spatial flows of each ecosystem sum to zero across the various compartments that make up this ecosystem, yielding a generalized source–sink constraint between ecosystem compartments.

In our simple metaecosystem with two ecosystems and two compartments, these constraints result in a *global material cycle* such that net flows at equilibrium are either in the direction $N_1 \rightarrow P_1 \rightarrow P_2 \rightarrow N_2 \rightarrow N_1$ (figure 7.9B) or in the opposite direction, depending on the sign of F_N^* (or any other function) in (7.20). In this global cycle, even though production and nutrient recycling occur within each ecosystem (figure 7.9A), one ecosystem acts as a net global producer ($N_1 \rightarrow P_1$), whereas the other acts as a net global recycler ($P_2 \rightarrow N_2$) (figure 7.9B). When there are more than two ecosystem compartments and local ecosystems, the pattern of material circulation in the metaecosystem may be more complex, but all local ecosystems are embedded in a web of material flows constrained by the functioning of the metaecosystem as a whole (Loreau et al. 2003b). In seasonal temperate systems, however, these constraints may change during the course of the year, such that each ecosystem may become alternatively source and sink for a given compartment (Nakano and Murakami 2001).

This simple, closed metaecosystem model shows that strong constraints on local ecosystem functioning emerge from spatial coupling of ecosystems. Although these constraints are likely to be less stringent in metaecosystems that are open to material exchanges with the outside world (and most real metaecosystems are open), they can nevertheless have significant impacts on local ecosystems. They imply that local ecosystems can no longer be governed by local interactions alone. Instead, by being part of the larger-scale metaecosystem, local ecosystems are constrained to become permanent or alternate sources and sinks for different compartments and thereby to fulfill different "functions" in the metaecosystem.

It is also conceivable, however, that these constraints may be impossible to meet in some cases because of unbalanced nutrient flows between ecosystems. During transient dynamics parts of the metaecosystem will then "absorb" others by progressively depriving them of the limiting nutrient. This means concretely that nutrient source–sink dynamics within metaecosystems may drive or accelerate successional changes until equilibrium is achieved and the final metaecosystem state becomes compatible with global source–sink constraints. This process of ecosystem absorption during

succession is reminiscent of Margalef's (1963) hypothesis that mature ecosystems such as forests "exploit" ecosystems from earlier successional
stages such as grasslands because animal consumers from late-successional
ecosystems move to nearby early-successional ecosystems for foraging.
Whether energy and material transfers across ecosystem boundaries are
strong enough to drive succession, however, depends on their magnitude
relative to that of the colonization processes that bring new species into
local ecosystems and thereby change their properties. This suggests that
combining an explicit accounting of spatial flows of energy and materials
with the dynamics of colonization of new patches by organisms in an integrated metaecosystem approach may provide a promising novel perspective on succession theory.

Although no empirical study has yet established complete energy or material budgets for coupled ecosystems, there are many examples of sizable
impacts of spatial flows on ecosystem functioning. In particular, significant reciprocal exchanges of materials and organisms occur at the interface between terrestrial and aquatic ecosystems. Secondary productivity in
rivers and lakes is often supported by inputs of litter and invertebrates
from nearby forests, while adult insects emerging from these lakes and
streams feed invertebrate and vertebrate predators from neighboring forests and grasslands. In a temperate forest-stream ecotone, reciprocal spatial flows of invertebrate prey alone accounted for 26 percent and 44 percent of the total annual energy budgets of forest birds and stream fishes,
respectively (Nakano and Murakami 2001). These spatial flows can also
have a cascade of indirect effects on recipient ecosystems. In particular,
they generally increase the top-down effects of predators on lower trophic
levels, which might explain differences across ecosystem types in the strength
of trophic cascades (Leroux and Loreau 2008). They can even generate
trophic cascades across ecosystems. Thus, fish can indirectly facilitate terrestrial plant reproduction through cascading trophic interactions across
ecosystem boundaries, by reducing the abundance of larval dragonflies in
ponds, thereby also reducing the abundance of adult dragonflies nearby,
which in turn increases the abundance and activity of insect pollinators
that are preyed upon by adult dragonflies (Knight et al. 2005).

CONCLUSION

Metacommunity and metaecosystem theory represents an important and
timely development for spatial ecology, a development that has the potential

to integrate the perspectives of community, ecosystem, and landscape ecology. At a time when humans are profoundly altering the structure and functioning of natural landscapes, understanding and predicting the consequences of these changes is critical for designing appropriate conservation and management strategies. Metacommunity and metaecosystem perspectives provide powerful tools to meet this goal. By explicitly considering the spatial interconnections among systems, they have the potential to provide novel insights into the dynamics and functioning of ecosystems from local to global scales, and to increase our ability to predict the consequences of land-use changes on biodiversity and the provision of ecosystem services to human societies.

In this chapter I have provided some examples of emergent properties that arise from spatial coupling of local ecosystems. These range from the coupled dynamics of local and regional diversity, through diversity–productivity relationships at local and regional scales, to patterns of nutrient flows from landscape to global scales. In all these examples, metacommunity or metaecosystem connectivity, as determined by the spatial arrangement of component ecosystems and the movements of organisms, energy, and inorganic substances across these ecosystems, exerts strong constraints on the structure, functioning, and stability of the system at both local and regional scales. It also drives many of the community and ecosystem properties that are traditionally studied at separate scales without consideration of these critical connections among scales. This shows that the metacommunity and metaecosystem perspectives offer a promising theoretical framework to explore hierarchical systems and emergent properties in a spatial context. Metaecosystems, however, have so far received much less attention than metacommunities. Developing our understanding of their properties and consequences is an important future challenge, which might well change the way we view the functioning of ecosystems and their management in the face of growing anthropogenic alterations of the Earth system.

Evolution of Ecosystems and Ecosystem Properties

Ecosystem ecology and evolutionary biology are two disciplines that have not had a history of close, peaceful relationships. They have been largely separate intellectual endeavors (Holt 1995), and when they have interacted, it has been more often to clash than to blend. The modern theory of evolution sees evolution as the result of a two-step process: trait variation is first generated at random by mutations or recombination of the genetic material, and natural selection then acts on this variation to sift out those traits that confer better adaptation to the environment. Since genes are carried, expressed, and transmitted by individual organisms, the individual organism is widely regarded as the main unit of selection, while the ecosystem in which the individual organism lives is viewed as part of the broad environmental context that determines the direction and strength of natural selection.

On the other hand, ecology is concerned with the multiple interactions that organisms have with their biotic and abiotic environment. As a discipline that studies the overall functioning of the systems made up of organisms and their environment, ecosystem ecology has had the natural tendency to view ecosystems as integrated units on their own and hence to search for laws and principles that govern the development and evolution of these higher-level units. Lotka (1922) had already proposed that, as a principle, "natural selection tends to make the energy flux through the system a maximum." His argument was simple but cogent: "If sources are presented, capable of supplying available energy in excess of that actually being tapped by the entire system of living organisms, then an opportunity is furnished for suitably constituted organisms to enlarge the total energy flux through the system." This idea was taken up by Odum in the form of his "maximum power principle" (H. T. Odum and Pinkerton 1955; H. T. Odum 1983), and by many others since in various guises. Other scientists have claimed that ecosystems evolve toward maximum entropy production, in agreement with the second law of thermodynamics (Schneider and Kay 1994). Yet others have argued that ecosystem stability leads to longer ecosystem persistence and hence should be favored by evolution on large time scales (Dunbar 1960). Fath et al. (2001) provide a nonexhaustive list of 10 different maximum principles that ecosystems are supposed to obey and examine how they are related to each other. Although many of these hypotheses are plausible and some of them are consistent with broad-scale ecological patterns, their main weakness lies in the fact that they are not explicitly connected to the evolutionary dynamics that takes place at the individual level, which is the main focus of the theory of evolution. As a result, evolutionary biologists have usually disregarded them as wishful thinking.

Perhaps the culmination of this mismatch between the perspectives of the two disciplines is the controversy over the Gaia hypothesis, which is still raging. Lovelock (1979), a geochemist, observed that Earth's atmosphere is in an extreme state of chemical disequilibrium and that this disequilibrium is fairly stable and maintained by large gas fluxes resulting from the activity of living organisms in the biosphere. Based on these uncontroversial observations, he proposed the more controversial hypothesis that the entire Earth system behaves as a sort of superorganism, Gaia, in which organisms collectively contribute to self-regulating feedback mechanisms that keep Earth's surface environment stable and habitable for life. Evolutionary biologists such as Dawkins (1982) opposed this hypothesis based on the argument that the Earth system is not a unit of selection, and hence there is no reason why evolution should lead to a planetary environment that is favorable for life. Watson and Lovelock (1983) then elaborated a hypothetical model of planetary regulation, Daisyworld, that incorporates natural selection among daisies with different colors to show that their hypothesis was consistent with evolutionary theory. A whole debate ensued about the relevance and limitations of this approach, which I do not intend to review here in any detail (see Lenton 1998; Free and Barton 2007 for reviews). Daisyworld is a metaphor, not a realistic representation of the Earth system. The point is that it has failed to resolve some of the fundamental issues related to the possible evolution of planetary regulation and hence to convince evolutionary biologists. As a result, the Gaia hypothesis remains a vivid example of the dialogue of the deaf that has often characterized the relationship between evolutionary biology and some branches of ecosystem ecology or biogeochemistry.

Yet an evolutionary ecosystem ecology surely must be possible since the properties that emerge at the ecosystem level, from the local to the global scale, are inextricably linked to the traits of the organisms that constitute ecosystems, and these traits are the result of evolution. The interplay

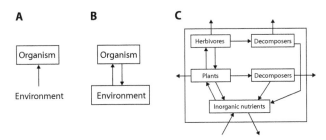

FIGURE 8.1. Three views on how natural selection operates: (A) the classical view, in which the environment is regarded as external to the organism and constant; (B) the modern view, which recognizes that organisms modify their environment, thereby generating an evolutionary feedback between the organism and its environment; and (C) the ecosystem view advocated here, which resolves an organism's environment into biotic and abiotic components that constitute the ecosystem. Modified from Loreau (2002).

between the ecological and evolutionary dynamics of communities and ecosystems is receiving increasing attention (Fussmann et al. 2007). The big question is: what is the nature of the link between evolution of species and ecosystem functioning?

Rather than attempting to answer this question straightaway, we can start by recognizing that the dichotomy between evolving organisms and the environment that determines the direction and speed of their evolution is not as clear-cut as assumed in classical evolutionary models and theories. In these models, the environment is regarded as external to the organism and constant (figure 8.1A). Although most evolutionary biologists today would probably agree that this assumption is an oversimplification of reality, it has been widely used, implicitly or explicitly, in evolutionary biology and population genetics because of its simplicity. In reality, organisms modify and interact with their environment in many different ways, which generates a feedback between the organism and its environment in the operation of natural selection (figure 8.1B). This environmental feedback may be mediated simply by interactions with conspecifics, but, more generally, it involves an active ecological role on the part of the organism, which transforms its biotic and abiotic environment (Lewontin 1983; Odling-Smee et al. 2003).

An organism's "environment", however, is still an abstract construct. This environment ultimately resolves into the concrete physical, chemical, and biological components with which the organism interacts. But this set of physical, chemical and biological components is precisely the ecosystem of which the organism is part (figure. 8.1C). Since each organism's environment is constituted by other organisms and abiotic factors, the ecosystem concept contains both the organisms and their environments and thus to some extent transcends the duality between organism and environment. Recognizing the ecosystem as the proper context within which natural selection, and hence evolution, operates is a major challenge for both ecology and evolution today.

Those willing to follow this path will no doubt encounter formidable obstacles. But these obstacles are a small price to pay for the big reward of contributing a more unified view of ecology and evolution. In this chapter, I shall follow this path with no pretense of reaching its end. I shall explore how evolution affects ecosystem functioning and how ecosystem functioning simultaneously affects evolution. By doing so without preconceived ideas, we shall discover that both ecosystem ecology and evolutionary biology provide key elements of a possible synthesis.

HOW CAN NATURAL SELECTION LEAD TO EVOLUTION OF ECOSYSTEM PROPERTIES?

At the root of the controversies that have plagued the relationships between ecosystem ecology and evolutionary biology lies the recurrent failure of ecosystem ecology to explicitly anchor its theories or hypotheses in the evolutionary mechanisms recognized by the modern theory of evolution, in particular, natural selection. Therefore, before examining the implications of the linkages between evolution and ecosystem functioning, I shall first discuss various ways in which natural selection can lead to evolution of ecosystem properties and the conditions under which each mechanism is likely to operate. This will hopefully avoid some of the past misunderstandings and controversies and allow a more rigorous approach to evolution of ecosystems and ecosystem properties.

Evolution of ecosystem properties can occur in at least three different ways, which require increasingly stringent ecological conditions (table 8.1). First, *evolution of individual species* though classical individual-level selection can indirectly drive changes in ecosystem properties. This simply requires that the species considered play an important role in the ecosystem, such that evolutionary changes in these species affect overall ecosystem functioning. Thus, in the next section, we shall see how evolutionary changes in the competitive abilities of either plants or decomposers can have considerable impacts on the productivity and nutrient cycling efficiency of ecosystems. These ecosystem-level impacts, however, are mere by-products

TABLE 8.1. Ecological Conditions Required for the Operation of Different
Modalities of Evolution of Ecosystem Properties

Modalities of evolution	Independent evolution	Coevolution
Evolution of species that affect ecosystem properties	Species plays a significant functional role	At least one species plays a significant functional role Long-lasting species interactions
Evolutionary feedback between organisms and ecosystem properties	Species plays a significant functional role Long-lasting species–environment interactions	At least one species plays a significant functional role Long-lasting species and species–environment interactions
Ecosystem-level selection of ecosystem properties	Species play significant functional roles Long-lasting, localized species and species–environment interactions	Species play significant functional roles Long-lasting, localized species and species–environment interactions

of the evolution of individual species and do not require feedbacks of eco-system functioning on evolution. Many of the examples provided by the emerging field of community genetics (Whitham et al. 2006) fall within this category because they consider the indirect effects of genetic changes in focal species such as dominant plants on their associated communities of insect herbivores or other organisms.

In a second scenario, evolution of individual species indirectly drives changes in ecosystem properties as before, but these changes in ecosystem properties in turn affect the evolution of the species considered through changes in their fitness. This establishes an *evolutionary feedback between organisms and ecosystem properties* similar to the situation depicted in fig-ure 8.1B and C, in which the species' environment includes ecosystem properties. This scenario requires not only that the species considered play an important role in the ecosystem but also that the species–ecosystem in-teraction be sufficiently long-lasting so that the evolutionary feedback is effective. It corresponds most closely to the concept of "niche construc-tion" (Odling-Smee et al. 2003), which attempts to capture the evolution-ary feedback between organisms and their environment. Ecosystem "engi-neers" (Jones et al. 1994), such as beavers building dams and earthworms modifying soil structure, are typical examples that fall within this category. The species involved, however, need not physically engineer the ecosystem. Later in this chapter, we shall consider a more complex example in which plants evolve resistance or tolerance to grazing depending on the indirect benefit they gain from nutrient cycling by herbivores.

In both of the above scenarios, an additional level of complexity may be introduced by *coevolution* of several interacting species. Species generally do not evolve in isolation or just in interaction with the ecosystem as a whole. They often establish stronger interactions with a few other species. If these interactions are maintained through time, they lead to coevolution of the interacting partners. If at least one of the coevolving species plays an important role in the ecosystem, coevolution can indirectly drive changes in ecosystem properties with or without an evolutionary feedback between species and ecosystem properties. The most complex situation occurs when there is diffuse coevolution among a large number of interacting species that collectively govern ecosystem processes, and these feed back on the coevolutionary process. We shall consider a theoretical example of this situation with the evolutionary emergence of complex food webs later in this chapter.

The final, and probably most controversial, potential mechanism leading to evolution of ecosystem properties is *ecosystem-level selection* of ecosystem properties. In this scenario, the fitness of the organisms that constitute the ecosystem is determined by their collective behavior, and natural selection operates among organisms from different ecosystems such that the best-performing species assemblages are selected. This scenario requires stringent ecological conditions. Not only does it require that the species involved play a significant role in the ecosystem and that species interactions be sufficiently long-lasting, it also requires that species interactions be strongly localized such that natural selection among local ecosystems can occur. I shall provide theoretical examples of this scenario with the evolution of plant–decomposer and plant–herbivore interactions in a heterogeneous environment later in this chapter. Again, coevolution may or may not occur in this scenario depending on the nature of species interactions. Although the examples that I shall discuss do not involve coevolution of plants and either decomposers or herbivores, coevolution is likely to occur under natural conditions that are conducive to ecosystem-level selection.

Although pure ecosystem-level selection is unlikely to occur in nature, it is a useful limiting case to consider because it lies at the end of a continuum between individual-level selection and group-level selection. *Multilevel selection theory* (Wilson 1980,1997; Sober and Wilson 1997) provides a consistent, unified, theoretical framework to address natural selection at

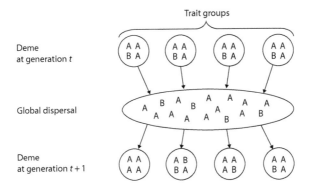

FIGURE 8.2. The spatial population structure assumed in Wilson's structured deme model.

multiple hierarchical levels. In this theory, the gene is the basic entity that is transmitted across generations, and hence it is the vehicle of natural selection, but the properties or traits that change in frequency across generations, and hence on which natural selection operates (Price 1995), can be expressed at any level of the biological hierarchy, from genes to ecosystems. In this framework, individuals are but one level of integration of genes among others. If some trait varies among groups and leads to differential proliferation or survival of these groups, the cause of natural selection is to be found at the group level, even though it is always possible in principle to reduce this phenomenon to properties of lower-level units, in particular, individual organisms and, ultimately, genes. The fact that selection of group properties can be reduced to individual selection by attributing a group's properties to the individuals that compose the group has led to a relatively sterile controversy about multilevel selection theory. Group selection is a useful concept to the extent that group properties have a significant effect on the selection of individuals' traits, just as individual selection is a useful concept because individual properties affect which genes are transmitted from one generation to the next.

Under what conditions is group selection expected to play a significant role in evolution? Wilson's (1980) structured deme model provides an elegant quantitative answer to this question. Assume that the life cycle of a haploid species includes two stages that take place at two different spatial scales: a stage in which most ecological interactions take place locally within a "trait group," and a reproductive stage in which global dispersal and reproduction take place within the "deme," i.e., the entire population of interbreeding individuals (figure 8.2). Assume also that a single trait determined by a single locus with two alleles, A and B, determines how an individual's fitness changes during its ecological interactions with its neighbors within a trait group. Allele A governs the expression of a trait that affects all individuals within the trait group, while allele B does not. Call an individual manifesting the trait (A type only) the donor, and all those affected by it (both A and B types) the recipients, and let a and b measure the additive effects of each donor on the fitness of the donor itself and on that of the recipients, respectively. If there are m A individuals in a trait group, the fitness changes of A and B individuals compared with their baseline fitness are then

$$f_A = a + (m - 1)b,$$
$$f_B = mb.$$
(8.1)

When there is a single trait group or when the deme is homogeneous such that all trait groups are identical, natural selection favors A if and only if $f_A > f_B$, i.e., from equations (8.1), if and only if

$$a > b.$$
(8.2)

This is the familiar result of individual selection: for a trait to be selected, it must give the donor a higher relative fitness than the recipients. Thus, "selfishness" is selected for.

If the deme is heterogeneous, however, such that different trait groups have different frequencies of A and B individuals, the condition for natural selection to favor A after global mixing of the deme is now that the average fitness change of A individuals be greater than that of B individuals; i.e.,

$$\overline{f_A} = \frac{\sum_{m,n} P_{mn} m[a + (m - 1)b]}{\sum_{m,n} P_{mn} m} > \overline{f_B} = \frac{\sum_{m,n} P_{mn} nmb}{\sum_{m,n} P_{mn} n},$$
(8.3)

where n is the number of B individuals in a trait group, and P_{mn} is the probability of having a trait group with m A individuals and n B individuals.

Using the properties of variances and covariances, inequality (8.3) can be rewritten as (Wilson 1980)

$$a > b\left[1 + \frac{\text{cov}(m,n)}{\overline{n}} - \frac{\sigma_m^2}{\overline{m}}\right].$$
(8.4)

Thus, inequality (8.2), which describes the outcome of natural selection within a single trait group or in a homogeneous deme, is a special case of the more general inequality (8.4), which describes natural selection in a heterogeneous deme. When there is no spatial variability in the population, the variances and covariances in inequality (8.4) vanish, leading to inequality (8.2).

Virtually no natural population, however, is spatially homogeneous. Any spatial variability generates a positive spatial variance, thus decreasing the right-hand side of inequality (8.4) and favoring the natural selection of A. Negative spatial covariances between A and B individuals further decrease the right-hand side of inequality (8.4) and favor the natural selection of A.

A useful case to consider is that when traits groups have a constant size $m + n = N$. Let $p = m/N$ and $q = n/N = 1 - p$ then be the proportions of A and B individuals, respectively, in a trait group. Noting that $cov(m,n) = -\sigma_m^2$, and $\sigma_m^2 = N^2\sigma_p^2$, inequality (8.4) becomes

$$a > b(1 - Ns_p^2), \qquad (8.5)$$

where

$$s_p^2 = \frac{\sigma_p^2}{\bar{p}\cdot\bar{q}} \qquad (8.6)$$

is a scaled spatial variance of the proportion of A individuals in trait groups. This scaled spatial variance varies between 0 and 1, and thus provides an appropriate standardized measure of spatial heterogeneity. The following limiting cases bracket the range of possible outcomes along a gradient of increasing spatial heterogeneity:

1. *Homogeneous deme*: In this case, $s_p^2 = 0$, yielding condition (8.2). As discussed above, this limiting case corresponds to pure individual selection (no variation, and hence no selection, between trait groups) and leads to evolution of "selfishness."

2. *Random distribution of types*: This yields a binomial distribution with $\sigma_m^2 = N\bar{p}.\bar{q}$ and hence $s_p^2 = 1/N$. Condition (8.5) for the selection of trait A then reduces to $a > 0$, which means that the donor must obtain a fitness benefit from the expression of trait A irrespective of the latter's effect on recipients. Thus, a random spatial distribution of types suffices to allow the natural selection of "weak altruism," i.e., of traits that are beneficial to the donor but may be even more beneficial to recipients.

3. *Maximally heterogeneous deme (monotypic trait groups)*: When each trait group is made up of a single type, $\sigma_p^2 = \bar{p^2} - \bar{p}^2 = \bar{p} - \bar{p}^2 = \bar{p}.\bar{q}$, and $s_p^2 = 1$. Condition (8.5) for the selection of trait A then becomes $a > b (1 - N)$. Since the right-hand side of this inequality is negative for any $N > 1$, evolution of "strong altruism," i.e., of traits that are detrimental to the donor but beneficial to recipients, is now possible. This limiting case corresponds to pure group selection (no individual selection within each trait group).

Multilevel selection theory in general, and Wilson's structured deme model in particular, predict a balance between individual selection for selfish traits within trait groups and group selection for altruistic traits between trait groups, which is governed by the level of spatial heterogeneity of the deme. As spatial heterogeneity increases, genetic relatedness within trait groups increases, and group selection becomes stronger, just as with kin selection. In fact, *kin selection* may be regarded as a special case of group selection in which genetic relatedness is generated by common descent. And reciprocally, group selection can be reanalyzed within the framework of kin selection theory provided the concept of kin is extended to include any form of genetic relatedness, whatever its origin. In particular, inequality (8.5) is equivalent to Hamilton's (1964a,b) classical rule stating that altruism is favored by natural selection if the direct fitness cost to the donor, c, is smaller than the benefit to the recipient, b, times its relatedness to the donor, r. Here, $c = -a$, and

$$r = \frac{Ns_p^2 - 1}{N - 1}. \tag{8.7}$$

Thus, relatedness varies from $-1/(N - 1)$ in a homogeneous deme (Gardner and West 2004) to $+1$ in a maximally heterogeneous deme. The only difference between inequality (8.5) and Hamilton's rule is that the former takes into account group size, and hence the number of recipients a donor interacts with, while the latter is expressed in terms of an average pairwise interaction between a donor and a recipient, hence the $N - 1$ term in the denominator of equation (8.7). Decreasing the size of trait groups generally has the same effect as increasing spatial heterogeneity because monotypic groups are more likely when groups have a small size.

It is important to notice that Wilson's structured deme model is just one elegant way to tackle multilevel selection in quantitative terms but that the basic concepts and processes involved in multilevel selection theory apply to many different forms of spatial structure. Spatial self-organization of ecological systems emerges spontaneously from local species interactions coupled with limited dispersal in a large number of models, even in continuous space. Although these self-organized spatial structures change through time, they provide a considerable amount of spatial variation on which community- or ecosystem-level selection can act (Johnson and Boerlijst 2002). Therefore, the theory of complex adaptive systems, which views system-level properties as emerging from localized interactions among individual agents (Levin 1999), is not incompatible with the operation of a significant, if implicit, amount of ecosystem-level selection. A broader synthesis of these different theoretical approaches would be particularly valuable.

PLANT–DECOMPOSER INTERACTIONS, AND EVOLUTION OF ECOSYSTEM PROPERTIES

I argued in chapter 6 that material cycling is a key process that imposes strong constraints on the overall functioning of ecosystems as well as on the dynamics and evolution of their component organisms. In this chapter, I want to explore some of the evolutionary consequences of material cycling.

To begin with, let us revisit the plant–decomposer ecosystem model (6.7) that was presented and discussed in chapter 6. We saw that two sets of critical parameters that are under the control of species traits and affect the ecosystem-level nutrient input–output balance are the resource competitive abilities and nutrient conservation efficiencies of plants and decomposers. Increasing the resource competitive ability of plants or decomposers increases ecosystem cycling efficiency and hence primary as well as secondary production (figure 6.3). Increasing their nutrient conservation efficiency has qualitatively similar effects (figure 6.4). Despite this similarity in their ecosystem-level effects, these two sets of parameters behave completely differently with respect to natural selection.

Resource competition theory (chapter 2) predicts that the plant species with the highest resource competitive ability (lowest N^*) outcompetes all other plant species; similarly, the decomposer species with the highest resource competitive ability (lowest M^*) outcompetes all the others. This rule applies to both ecological and evolutionary time scales and appeals to competition between types within a homogeneous material cycle, i.e., to individual selection. Thus, *within-cycle competition* is a force that spontaneously leads to more materially closed, productive ecosystems. In this case, evolution of ecosystem properties is a mere by-product of the evolution of individual organisms that use nonliving resources in ecosystems.

In contrast, nutrient conservation efficiency is a trait that is selectively neutral within a homogeneous material cycle because it does not affect the resource competitive ability of either plants or decomposers. Although nutrient conservation is a strategy that is strongly beneficial to all ecosystem components, the individual plants or decomposers that would express this trait would not derive any fitness benefit from it relative to their competitors. Therefore, individual selection cannot select for such traits. If these traits were costly, they would even be counterselected because they would amount to altruistic traits. Multilevel selection theory then predicts that sufficient spatial heterogeneity between material cycles is necessary to allow these traits to evolve through group or ecosystem selection.

Can ecosystem selection be reasonably expected in plant–decomposer systems? It is hard to tell in the absence of empirical data, but there are some indications that some degree of ecosystem selection may not be entirely unrealistic for these systems. There is growing evidence that a significant fraction of nutrient cycling may take place at much smaller spatial and temporal scales than previously believed. For instance, about two-thirds of nitrogen uptake by grasses originates from rapid mineralization of dead roots within their rooting system in some tropical savannas (Abbadie et al. 1992). These grasses even control nitrification in their immediate vicinity through a balance between inhibitory and stimulatory effects on nitrifying bacteria (Lata et al. 2000, 2004). In this case, a relatively tight association between individual plants and microbial populations should be expected. Such a strongly localized spatial structure tends to generate competition between organisms involved in spatially distinct cycles, a process that I called *between-cycle competition* (Loreau 1998b).

As an extreme case of between-cycle competition, consider a perfectly structured environment in which each individual plant occupies an isolated site during its lifetime and is associated with a single decomposer individual or colony. Assume that sites become vacant when previous occupants are extirpated by natural death or disturbance, and establishment of both plants and decomposers at vacant sites obeys a competitive lottery. Finally, assume that the probability of a genotype's successful establishment at a site is proportional to its total production in all other sites because higher production means production of more propagules of a higher quality.

The dynamics of site occupancy by plants then obeys the equation (Loreau 1998b)

$$\frac{dp_{Pi}}{dt} = p_{Pi}(r_{Pi}V_P - m_{Pi}), \tag{8.8}$$

where

$$r_{Pi} = \alpha_{Pi}\left(\frac{\sum_j \Phi_{Pij}}{p_{Pi}T}\right), \tag{8.9}$$

$$V_P = 1 - \sum_k p_{Pk}. \tag{8.10}$$

In this equation, T is the total number of sites available, p_{Pi} the proportion of sites occupied by plant genotype i, m_{Pi} its mortality rate, Φ_{Pij} its productivity at site j, and α_{Pi} its reproductive efficiency, a constant of proportionality that incorporates both the allocation of plant genotype i's production to reproduction and its ability to disperse and establish at new sites. The aggregated parameter r_{Pi}, which is plant genotype i's average productivity times its reproductive efficiency, represents a potential reproduction rate, reproduction here being considered completed after the establishment of offspring at new sites. Last, V_p is the proportion of vacant sites; only dispersal to vacant sites leads to successful reproduction. Equation (8.8) is a straightforward extension of the competitive lottery model (2.17) presented in chapter 2.

An equivalent equation holds for decomposers with a mere change in subscripts:

$$\frac{dp_{Di}}{dt} = p_{Di}(r_{Di}V_D - m_{Di}). \tag{8.11}$$

The outcome of this competition for vacant sites between species or genotypes is identical to that of the simple competitive lottery considered in chapter 2. At equilibrium, the fraction of vacant sites, V_X^*, must satisfy

$$V_X^* = m_{Xi} / r_{Xi}, \tag{8.12}$$

where $X = P$ or D. This relation can be satisfied only by a single species or genotype. Therefore, V_X^* here plays the same role as R^* in classical resource competition, and the species or genotype the lowest V_X^*, and hence the highest basic reproductive capacity (the inverse of V_X^*), eventually displaces all the others.

In the simplest case where plants and decomposers disperse independently and their effects on their local environment are additive, the outcome of this dual selective process is the selection of the material cycle that combines the plant and decomposer genotypes with the highest basic reproductive capacities. Now the basic reproductive capacity of a genotype is proportional to its average productivity at a site [equation (8.9)], everything else being equal. If the dynamics of site occupancy is slow compared with the dynamics of material cycles within sites, the latter will approach their equilibrium, so that traits that contribute to increase equilibrium productivities may be selected for. In particular, selection for increased nutrient conservation is possible, leading to enhanced ecosystem properties, namely, increased ecosystem cycling efficiency and primary and secondary productivities.

A noteworthy feature of this model is that material cycles within sites behave very much like *superorganisms* (Wilson and Sober 1989), where genotypes play the role of alleles at the plant and decomposer "loci" and the basic reproductive capacity is the measure of fitness. Like organisms, these cycles have a temporary existence, their properties result from the random assortment of their constituent genotypes, and the unit of selection is the entire genotype. Selection of traits advantageous to the cycle is then just as natural as selection of traits advantageous to the individual organism in classical individual selection theory. Unlike organisms, however, the biotic components of the material cycle reproduce separately.

Of course, the above scenario is extreme because it considers material cycles that have minimal size, thereby minimizing the effects of individual selection within cycles and maximizing the effects of ecosystem selection between cycles. In the case of plant–decomposer interactions with a single limiting nutrient, however, the two levels of competition or selection converge to yield similar outcomes at the ecosystem level, i.e., increased cycling efficiency, primary productivity, and secondary productivity. Although the two levels of competition or selection yield these outcomes indirectly through different pathways (within-cycle competition maximizes the resource-use intensity of plants and decomposers, while between-cycle competition maximizes their basic reproductive capacity), and these pathways may involve different constraints and trade-offs, the fact that they have convergent ecosystem-level effects precludes strong conflicts between them. To examine the consequences of conflicts between levels of selection for the evolution of ecosystem properties, I turn to the more conflictual plant–herbivore interactions in the next section.

PLANT–HERBIVORE INTERACTIONS, AND EVOLUTION OF ECOSYSTEM PROPERTIES

In chapter 6, I also analyzed the ecological effects of herbivores on plants within ecosystems. I showed that, although herbivores have a negative direct effect on plants through consumption of plant tissue, they have a positive indirect effect through nutrient cycling, and this positive indirect effect can outweigh the negative direct effect under specific, well-identified conditions. Does this imply that ecosystem-level constraints make the plant–herbivore interaction mutualistic, not antagonistic (Owen and Wiegert 1976, 1981)?

The evolutionary consequences of grazing optimization, and of ecological indirect interactions in general, are complex for two main reasons. First, increased plant productivity does not necessarily translate into increased plant fitness. It is still unclear which plant traits determine fitness. If a plant's seed production or other measures of fitness are mainly determined by its biomass, then no mutualistic interaction with herbivores is possible because plant consumption by herbivory always decreases plant biomass (chapter 6). On the other hand, if a plant's fitness is mainly determined by its productivity, then herbivory can increase plant fitness through increased productivity. Reality probably lies between these two extremes, and thus we may expect herbivory to increase plant fitness in some cases. Second, when it does, it is not absolute but relative fitness that counts. If two plant types (species or genotypes) are mixed, one of them being tolerant (mutualistic) and the other resistant (antagonistic) to herbivory, the resistant type is expected to outcompete the tolerant type because it benefits from the positive indirect effect of increased nutrient cycling but does not suffer the negative direct effect of herbivore consumption. As a result, the fitness of the resistant type is higher than that of the tolerant type, and tolerance does not evolve even though it is indirectly beneficial to both types. This evolutionary argument has been used by some to pronounce a death sentence for the idea of plant–herbivore indirect mutualism, indeed for any evolved indirect interaction (Belsky et al. 1993).

This argument, however, ignores two important factors that counteract the advantage of antiherbivore defense: spatial heterogeneity and the cost of defense. The *spatial structure* of the plant–herbivore system can generate spatially heterogeneous nutrient cycling, just as in the plant–decomposer interaction above. If herbivores recycle nutrient in the vicinity of the grazed plants, or if plants from the same type are aggregated, herbivores tend to recycle proportionally more nutrient on the plants that are grazed more heavily, thus augmenting the indirect benefit of grazing for the grazed plants. To examine this issue quantitatively, Claire de Mazancourt and I built a model that includes three nested spatial scales: individual plants, patches, and landscape (de Mazancourt and Loreau 2000b) (figure 8.3). Plants can be of two types, which differ in their resistance or tolerance to grazing by herbivores. They are distributed in a landscape among a number of discrete patches. Each of these patches contains a number of plants of both types. Landscape structure is generated by the interaction between plants and herbivores. On a small, within-patch spatial scale, herbivores recycle the nutrient limiting plant growth homogeneously without discriminating among plant types. On a large, between-patch spatial scale, herbivores have an ideal free distribution; i.e., they are distributed proportionally to the amount of food they obtain in a patch, and they recycle the limiting nutrient in the same proportion. In addition, each individual plant has access to a distinct nutrient depletion zone within a patch to maintain short-term coexistence. On a short time scale, plants reach an equilibrium primary production within each patch. On a longer time scale, however, they compete for space made vacant by mortality in a similar manner as in the above plant–decomposer model; i.e., the seed output of each plant type is proportional to its net primary production across the landscape.

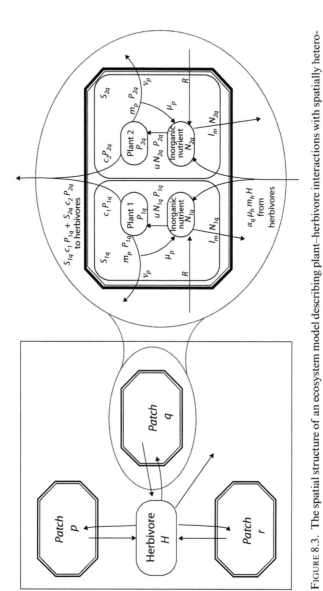

FIGURE 8.3. The spatial structure of an ecosystem model describing plant–herbivore interactions with spatially hetero-geneous nutrient cycling. Reprinted from de Mazancourt and Loreau (2000b).

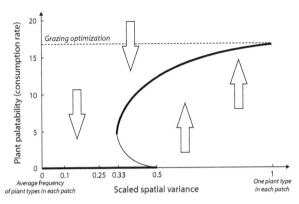

FIGURE 8.4. Evolutionary convergence stable strategies of plant palatability (in bold) vs. the spatial scaled variance of plant types [equation (8.6)] in the model depicted in figure 8.3. Arrows show the direction of selection given by the sign of the fitness gradient. The horizontal dashed line shows the level of plant palatability that yields grazing optimization. Modified from de Mazancourt and Loreau (2000b).

These assumptions provide all the ingredients necessary for the application of multilevel selection theory. There is a *conflict between levels of selection*: individual selection within patches favors the resistant type over the tolerant one because it has a higher relative fitness for the reason discussed above, whereas group selection between patches favors patches with a higher proportion of the tolerant type because they have a higher average absolute fitness and hence contribute more offspring to the deme as a whole. Evolution is then governed by the balance between these two conflicting levels of selection, just as in the evolution of altruism. Its outcome depends on the strength of spatial aggregation and patch size: tolerance to grazing evolves provided that spatial aggregation is strong enough or patch size is small enough. Both these factors contribute to increase within-patch plant relatedness, as defined by equation (8.7), or scaled spatial variance, as defined by equation (8.6), and hence the relative strength of group selection. Interestingly enough, the relationship between the evolutionary stable level of plant palatability (which determines the herbivore grazing rate) and scaled spatial variance proves to be highly nonlinear. Low levels of scaled spatial variance select for entirely nonpalatable plants, while high enough levels of scaled spatial variance abruptly switch the system toward selection of highly palatable plants, and two alternative evolutionary stable strategies coexist for intermediate values (figure 8.4). Thus, strong spatial heterogeneity or small patch size is necessary to foster evolution of tolerant

plants, which suggests that this mechanism might work only with small herbivores, such as phytophagous insects, which recycle nutrients in the immediate vicinity of the plants they consume. When this condition is met, plant
tolerance can easily evolve to high levels. Note, however, that the optimal
level of plant tolerance that yields grazing optimization is never reached
through evolution except under the most extreme scenario of pure group
selection. The presence of individual selection prevents the ecosystem from
reaching optimal functioning. I shall discuss the implications of this important conclusion later in this chapter.

The second factor that counteracts the advantage of antiherbivore defense is the *cost of defense* in terms of nutrient investment. This cost generates a trade-off in plants between antiherbivore defense and growth or nutrient uptake. To analyze its effects on the evolution of plant defense, we
built a model similar to that described in figure 8.3 but with spatially ho-
mogenous nutrient recycling by herbivores to separate the effects of the cost
of defense from those of spatial heterogeneity (de Mazancourt et al. 2001).
For many ecologically plausible trade-offs, plant evolution leads to a single
evolutionary continuously stable strategy (CSS), i.e., a strategy to which
evolution converges and which cannot be invaded by any other close strategy. This CSS has complex relationships with the strategies that maximize
plant production or plant biomass, depending on ecosystem parameters.
Because of this complexity, different ecological and evolutionary scenarios
of herbivore addition or removal are possible, which highlight the ambigu-
ity of the notion of mutualism as it has been used by different authors.

It is useful to distinguish two types of mutualism: an *ecological* or *proximate mutualism* in which each species gains a benefit from the presence of
its partner in the absence of any evolutionary change, as revealed, e.g., by
an ecological press perturbation, and an *evolutionary* or *ultimate mutualism*, in which the mutual benefit persists even after evolution has occurred
(de Mazancourt et al. 2005). The conditions for an evolutionary mutualism are more stringent than those for an ecological mutualism because
each species may have *evolved dependence* upon its partner, so that the removal of one species may have a negative impact on the other in the short
term, but this negative impact may disappear after each species has had the
opportunity to evolve and adapt to the new conditions created by the absence of its partner (figure 8.5). For instance, a plant may adapt to the regular seasonal occurrence of herbivores by delaying its growth or reproductive
effort until grazing ceases. This adaptation may provide it with a competitive advantage over other plants, but it is likely to turn into a competitive
disadvantage if herbivores disappear for whatever reason. In this case, the

presence of herbivores is beneficial to the plant under the prevailing conditions that have led to the evolution of its adaptation (ecological or proximate mutualism), but the plant might be better off altogether without herbivores if it could get rid of this adaptation and evolve toward faster growth
and reproduction (evolved dependence but no evolutionary or ultimate
mutualism). An evolutionary or ultimate mutualism would require that the
plant be better off in the presence of herbivores and after evolution in their
presence than in the absence of herbivores and after evolution in their absence. Such a comparison, however, is difficult to perform, so that the ecological definition of mutualism seems most appropriate operationally (de
Mazancourt et al. 2005).

When a plant's reproductive ability is determined by its productivity,
herbivory is capable of improving plant performance on both an ecological
and an evolutionary time scale provided that herbivore recycling efficiency
is sufficiently greater than plant recycling efficiency, thus generating a
plant–herbivore mutualistic interaction (figure 8.6). As herbivore recycling
efficiency is increased, the plant–herbivore interaction becomes increasingly mutualistic—first ecologically, then evolutionarily. Counterintuitively,
however, plants simultaneously evolve to increase their level of antiherbivore defense (figure 8.7). This seemingly paradoxical result is explained by
the fact that defended plants gain a higher benefit from not being consumed relative to less defended plants as herbivore recycling increases. It
shows that mutualism can go hand in hand with increased conflict between
partners. Evolutionary conflicts between partners are known in other mutualistic interactions (Anstett et al. 1997; Law and Dieckmann 1998). Once
more, this example emphasizes the fact that individual selection generally
yields suboptimal behavior at the ecosystem level.

Experimental studies have demonstrated that under some conditions
herbivory leads to increased plant fitness, thus providing evidence for an
ecological plant–herbivore mutualism. For example, grazing or clipping
increased lifetime reproductive output in some populations of *Ipomopsis
aggregata* (Paige and Whitham 1987) and *Gentianella campestris* (Lennartsson et al. 1997) (figure 8.8). Evolved dependence, however, may well
explain these proximate benefits of herbivory (de Mazancourt et al. 2005).
Plants adapted to herbivory often develop mechanisms for resource mobilization triggered by herbivory to reduce herbivore damage. As in the above
hypothetical example, these adaptations lead to low reproduction in the
absence of herbivory because such plants are poorly adapted to a situation
without herbivores. In *G. campestris*, for instance, the ability to produce
more seeds after herbivore damage was only found in populations adapted

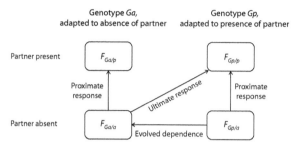

FIGURE 8.5. Definitions of ecological or proximate mutualism, of evolutionary or ultimate mutualism, and of evolved dependence based on the differences between the performances of two types of genotypes in two types of environments: a genotype adapted to the absence of its partner (left column) and a genotype adapted to the presence of its partner (right column), either in the presence (upper row) or in the absence (lower row) of its partner. The proximate response of the organism to partner removal is measured as $F_{Ga/p} - F_{Ga/a}$ for the genotype adapted to the partner, and as $F_{Gp/p} - F_{Gp/a}$ for the genotype adapted to the absence of the partner. The ultimate response of the organism to partner removal is measured as $F_{Gp/p} - F_{Ga/a}$, i.e., as the difference between the performance in the presence of the partner of a genotype that evolved with the partner and the performance in the absence of the partner of a genotype that evolved without the partner. Evolved dependence is measured as the difference between the performance without the partner of a genotype that evolved without the partner and the performance without the partner of a genotype that evolved with the partner, $F_{Ga/a} - F_{Gp/a}$. Note that the ultimate response of a genotype adapted to the partner equals its proximate response minus evolved dependence. Modified from de Mazancourt et al. (2005).

to either herbivory or mowing (figure 8.8). But I know of no empirical evidence for an ultimate benefit of herbivory to plants. Järemo et al. (1999) measured the ultimate response of *G. campestris* to herbivore removal using comparative data on populations that have evolved with and without herbivores and detected neither an ultimate benefit nor an ultimate cost, suggesting that the ultimate response of this plant species to herbivores is neutral.

EVOLUTIONARY EMERGENCE OF ECOSYSTEMS

Ecological theories and most evolutionary theories deal with systems that have given, preexisting structure and component species. But where do this structure and these component species come from in the first place? How do complex ecosystems emerge through evolution and how does this evolutionary history constrain present-day ecosystems? The generation of species diversity is a relatively classical issue in evolutionary ecology, but it has generally been circumscribed within the context of competitive communities in which species have similar environmental requirements and effects. By contrast, the generation of entire food webs or ecosystems is a novel issue that has begun to be explored theoretically only during the past few years.

Pioneering work on this topic was performed by McKane and colleagues (Caldarelli et al. 1998; Drossel et al. 2001; McKane 2004), who built the Webworld model, a complex simulation model in which species are represented by a vector of arbitrary binary traits that define their population dynamics and trophic interactions with other species. These traits are subject to mutations, which creates the potential for speciation events. Species may also go extinct as a result of species interactions. Their model was able to generate complex food webs that are reasonably similar to real food webs.

Since then, there has been a burst of *community evolution models*, many of which have been developed simultaneously and independently to explore the evolutionary emergence of food webs or more abstract interaction webs (Christensen et al. 2002; Loeuille and Loreau 2005; Ito and Ikegami 2006; Rossberg et al. 2006; Bell 2007). The main difference between these models, which determines their basic features and results, lies in the number and identity of traits they consider. Like the Webworld model, most existing community evolution models involve a large number of traits (Christensen et al. 2002; Rossberg et al. 2006; Bell 2007). This makes them flexible but also relatively abstract and difficult to test empirically. Ito and Ikegami (2006) built a continuous version of the Webworld model that includes only two traits for each species, one that describes it as a prey and another that describes it as a predator. Nicolas Loeuille and I built an even simpler food-web evolution model in which a single evolving trait, body size, determines each species' population dynamics and interactions (Loeuille and Loreau 2005).

The advantages of the latter approach are that it clearly identifies a measurable trait and the ecological trade-offs it generates, and as a result it makes empirically testable predictions. Although the ecology of a species obviously depends on more than a single trait, the number of traits considered is traded off in the models against the biological realism introduced in these traits. Body size is well known to play a key role in the physiological and ecological characteristics of species, and most of its effects have been quantified (Kleiber 1961; Peters 1983; Brown et al. 2004). Therefore, body

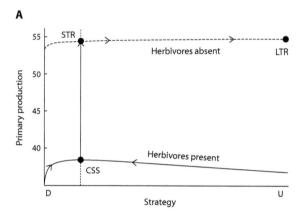

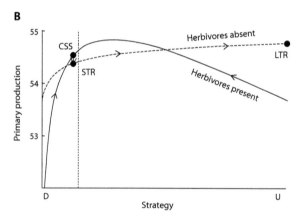

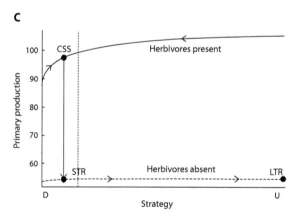

FIGURE 8.6. Primary production along the trade-off curve of plant strategies from fully defended (D) to fully undefended (U), in either the presence (solid lines) or the absence (dashed lines) of herbivores in a model of plant

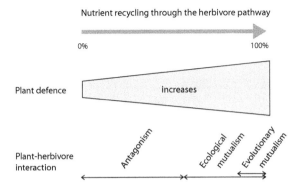

FIGURE 8.7. The paradox of evolution of plant antiherbivore defence: as herbivore recycling efficiency is increased, the plant-herbivore interaction becomes increasingly mutualistic (first ecologically, then evolutionarily), but
plants evolve to increase their level of defense.

size is a particularly appropriate trait to consider as a first step toward
building a testable theory of food-web evolution.

Our model is disarmingly simple. Ecological dynamics is described by a
set of two differential equations:

$$\frac{dN_i}{dt} = N_i\left[f(x_i)\sum_{j=0}^{i-1}\gamma(x_i - x_j)N_j - m(x_i) - \sum_{j=1}^{S}\alpha(x_i - x_j)N_j - \sum_{j=i+1}^{S}\gamma(x_j - x_i)N_j\right],$$

(8.13)

evolution with a cost to antiherbivore defense. When herbivores are present,
evolution leads to a continuously stable strategy (CSS). To test whether plant–
herbivore interactions are mutualistic, plant performance is compared in
the presence and in the absence of herbivores. Two responses of plants are
distinguished (figure 8.5): their proximate or short-term response (STR), in
which herbivores have been removed but plants have not yet adapted to the
herbivore-free situation, and an ultimate or long-term response (LTR), in
which herbivores have been removed and plants have had time to adapt.

(A) No mutualism: plant performance is always decreased in the presence
of herbivores. (B) Proximate or ecological mutualism: plant performance is
decreased by the short-term response to herbivore removal (STR is lower than
CSS). However, in the long term, the removal leads to an increase in primary
production (LTR is higher than CSS), and there is no evolutionary mutualism. (C) Ultimate or evolutionary mutualism: herbivore removal results in decreased plant performance in both the short and the long term (CSS higher
than STR and LTR).

The dotted line shows the CSS in case (A) for comparison with the other
panels. Modified from de Mazancourt et al. (2001).

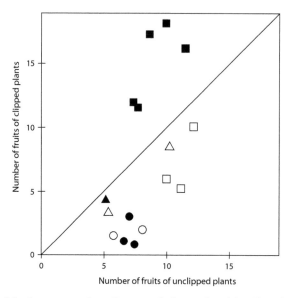

FIGURE 8.8. Average number of mature fruits produced by clipped and unclipped plants in the biennial grassland herb *Gentianella campestris*. At the diagonal compensation line, clipped and unclipped plants accomplish equal fruit production. Clipped plants overcompensate in populations located above the compensation line. The data include 10 populations that were early- (circles), intermediate- (triangles), or late-flowering (squares) and two populations with early-, intermediate-, and late-flowering subpopulations. The populations were sampled from grazed or mown (solid symbols) or ungrazed and unmown (open symbols) localities in central-eastern Sweden. Modified from Lennartson et al. (1997).

$$\frac{dN_0}{dt} = I - qN_0 - \sum_{i=1}^{S}\gamma(x_i)N_0N_i$$
$$+ (1 - \lambda)\left\{\sum_{i=1}^{S}m(x_i)N_i + \sum_{i=1}^{S}\sum_{j=0}^{i-1}[1 - f(x_i)]\gamma(x_i - x_j)N_iN_j\right.$$
$$\left. + \sum_{i=1}^{S}\sum_{j=1}^{S}\alpha(x_i - x_j)N_iN_j\right\}. \tag{8.14}$$

The first equation describes the dynamics of the biomass or nutrient stock, N_i, of any of the S morphs or species i in the ecosystem, while the second describes the dynamics of the inorganic stock of a limiting nutrient. Although the focus was on the evolutionary emergence of food-web structure, this is a full ecosystem model that satisfies the principle of mass conservation. The ecological characteristics of each species i are governed by its body mass, x_i. Its production efficiency, $f(x_i)$, and mass-specific mortality rate, $m(x_i)$, are assumed to follow the classical allometric relationships $f(x_i) = f_0 x_i^{-1/4}$ and $m(x_i) = m_0 x_i^{-1/4}$ (Kleiber 1961; Peters 1983; Brown et al. 2004). Since predators often consume prey that are smaller than they are but not too small, the consumption rate of prey species j by predator species i is assumed to be a Gaussian function of the difference between their body masses:

$$\gamma(x_i - x_j) = \frac{\gamma_0}{\sigma\sqrt{2\pi}} \exp\left[\frac{-(x_i - x_j - d)^2}{\sigma^2}\right], \tag{8.15}$$

where σ^2 is the variance of the consumption rate, and predators of size x_i forage optimally on prey of size $x_i - d$. Species that have similar body sizes are also more likely to use their habitat in similar ways and hence to interfere with each other. Interference competition between two species occurs at a rate that depends on the absolute difference between their body masses:

$$\alpha(x_i - x_j) = \begin{array}{ll} \alpha_0 & \text{if } |x_i - x_j| \leq \beta, \\ 0 & \text{if } |x_i - x_j| > \beta. \end{array} \tag{8.16}$$

Last, all the amounts of nutrient that are released during the processes of mortality, consumption, and interference are recycled in inorganic form with a probability $1 - \lambda$. The ecosystem also receives a constant input of inorganic nutrient, I, and loses inorganic nutrient at a rate q. Nutrient in inorganic form is assumed to have an arbitrary body mass of zero. Evolutionary dynamics is generated by mutations in body size, which occur at a rate of 10^{-6} per unit mass at each time step, and by species extinctions driven by ecological dynamics.

This relatively simple model generates complex food webs that emerge by evolution from a single ancestor through a succession of species replacement, coexistence, diversification, and divergence processes (figure 8.9). Diversification is very fast in the beginning, but the food web gradually stabilizes into an evolutionary quasiequilibrium in which species continue to turn over but food-web structure is relatively stable. These features are found in all community evolution models.

The final structure of the food web depends on model parameters. The dimensionality of the food web (total number of species and length of the food chain) is mainly determined by parameters that govern ecosystem productivity, such as the nutrient input, I, and the basal production efficiency, f_0. The shape of the food web, however, is mainly determined by two parameters that govern species interactions, i.e., the interference competition rate, α_0, and the width of the predators' food niche, $nw = \sigma^2/d$.

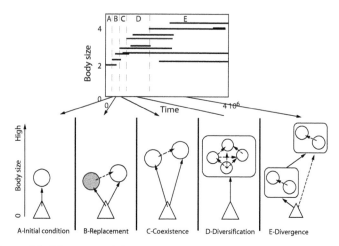

FIGURE 8.9. First steps of the emergence of a size-structured food web. The upper panel shows the trait composition of the community through time, while the lower panel details the different steps of the emergence. The simulation starts with a single species that consumes inorganic nutrient (panel A). Once in a while, mutants appear (here, larger than the resident) and replace their parent (panel B, in which the gray morph goes to extinction). After several replacements, an evolutionary branching happens, leading to coexistence of the mutant and the resident (panel C). A rapid diversification then occurs in which several morphs are able to coexist (panel D). These morphs are then selected to yield differentiated trophic levels (panel E). Reprinted from Loeuille and Loreau (2010).

When there is no interference competition, diversity within a trophic level is reduced, and the food web tends to reduce to a simple food chain. Weak interference, however, suffices to generate very diverse food webs. At the other extreme, if interference is very strong, individual fitness is mostly determined by competition, while selective pressures due to trophic interactions are less important. As a result, species body sizes become evenly spaced, and the distinction between trophic levels disappears. Food niche width determines species' degree of specialization. When food niches are narrow, distinct trophic levels separated by a difference d in body size emerge because it is selectively advantageous to have a body size d to efficiently use the inorganic resource (which has zero body size), a body size $2d$ to efficiently use the first trophic level, and so on. By contrast, when food niches are wide, the selective advantages of having a body size d, $2d$, etc., are smaller, which leads to more omnivorous diets and blurred trophic levels. The interplay of these two parameters is able to produce a complete continuum of food-web structures (figure 8.10).

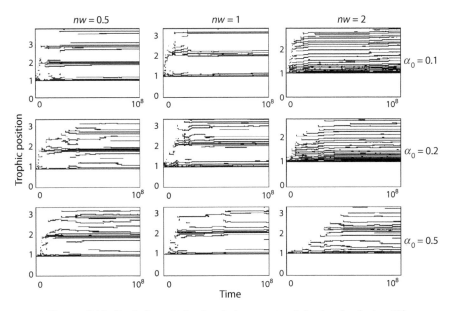

FIGURE 8.10. Evolution of simulated size-structured food webs during 10^8 time steps for three values of niche width (*nw*) and competition intensity (α_0). Trophic position is determined recursively from the bottom to the top of the food web. The trophic position of a target species is defined as the average trophic position of the species it consumes weighted by the proportion of nutrient these represent in the target species' diet, plus 1. Since this measure is strongly correlated with body size, similar patterns are obtained using body size. Reprinted from Loeuille and Loreau (2005).

A striking and unexpected feature of our evolutionary model is that it is able to fit empirical data on food-web structure, such as connectance, food chain length, proportion of omnivores, and proportions of basal, intermediate, and top species, just as well as, or even better than, traditional static food-web models such as the cascade model (Cohen et al. 1990), the niche model (Williams and Martinez 2000), and the nested hierarchy model (Cattin et al. 2004). This is again a general feature of food-web evolution models (Caldarelli et al. 1998; Loeuille and Loreau 2005; Rossberg et al. 2006; Bell 2007). The first conclusion that can be drawn from this observation is that mere data fitting, as performed traditionally, is a poor test of model performance. The same conclusion is valid for many areas of ecology. For instance, the ability of a theory to fit empirical species abundance distributions is poor evidence for its usefulness since a wide range of models can do this (McGill et al. 2007).

More interestingly, our evolutionary model is able to fit empirical data on large-scale food-web attributes by varying only two parameters that govern small-scale species interactions, i.e., interference rate and food niche width (Loeuille and Loreau 2005). This is a significant improvement over static food-web models, which use large-scale attributes such as species diversity and connectance, to predict other large-sale attributes but which cannot account for species diversity and connectance in the first place. It is also an improvement over other community evolution models, which use a large number of poorly specified traits. Interference and food niche width are two individual-level traits that are easy to define and measure. This makes the model's theoretical predictions testable.

Clearly, our evolutionary model has limitations. In particular, body size is not the sole trait determining a species' ecology; body size governs large-scale trends in species properties across orders-of-magnitude variations, but other traits determine large residual variations around these trends. Also, our model does not account for the basic functional differences between autotrophs, consumers, and decomposers. As a consequence, it applies mainly to the animal consumer part of food webs. The evolutionary dynamics it represents are essentially clonal and do not include many of the complexities that arise with sexuality in the speciation process and in the maintenance of intraspecific genetic variation. Properties at the level of the ecosystem as a whole have been poorly investigated so far. There is still room for vast improvements and exploration of a wide range of topics.

While the evolution of food webs begins to be explored thoroughly using a variety of models, the evolution of complete ecosystems, including their autotroph and decomposer components, remains a future challenge. Downing and Zvirinski (1999), Downing (2002), and Williams and Lenton (2007) have taken promising steps in that direction by exploring the evolutionary emergence of nutrient cycling in microbial systems. Their models simulate evolving communities of microorganisms in which traits that govern nutrient uptake and release, as well as effects on and response to other environmental variables, are subject to mutation and selection. They typically lead to the emergence of a diversity of biochemical guilds with complementary nutrient uptake and release patterns, which collectively recycle nutrients, sustain high biomass, and regulate their abiotic environment. Although they have been conceived within the framework of Gaia theory to explore the evolution of global environmental regulation—a feature that may not be the most robust because of the potential for "rebel" organisms that shift the environment away from the state to which the majority of the community is adapted (Williams and Lenton 2007)—these models provide stimulating insights into the evolution of nutrient cycling and functional complementarity between biochemical guilds through nontrophic interactions.

Community evolution models make three important contributions to community and ecosystem ecology (Loeuille and Loreau, 2010). First, they extend classical pairwise coevolutionary models to large, complex ecosystems with new results. For example, coevolution in simple, two-species communities can have destabilizing effects on population dynamics. By contrast, community evolution models suggest that evolution tends to produce large, complex communities that are relatively stable. Second, they provide, for the first time, insights into the evolutionary emergence of entire food webs or ecosystems. Classical evolutionary models have mostly considered evolution or coevolution of preexisting species. In community evolution models, species themselves emerge spontaneously from the evolutionary dynamics of the system. Third, they provide new perspectives on community and ecosystem properties and potentially a better understanding of the mechanisms that generate them. They describe species interactions based on individual-level traits, so that community and ecosystem properties are emergent properties of processes that take place at a smaller scale. Despite their current limitations, community evolution models have promising prospects. I view them as a first step toward building an evolutionary ecology of complex ecosystems. They now need to be taken one step further and made more realistic to deliver testable predictions on the overall structure and functioning of ecosystems.

EVOLUTION, ENTROPY MAXIMIZATION, AND MAXIMUM PRINCIPLES IN ECOLOGY

To end this chapter I would like to reexamine and discuss one of the most contentious issues in ecology, that is, the existence of *maximum (or minimum) principles* at the ecosystem level. As I briefly mentioned in the introduction to this chapter, the idea that ecosystems develop or evolve toward maximization (or minimization) of some ecosystem properties is a longstanding idea. Many different ecosystem properties have been proposed as candidates for maximization during succession or evolution. Fath et al. (2001, 2004) reviewed a number of these maximum principles, their connections, and complementarity. They summed them up in the following maxim: "Get as much as you can (maximize input and first-passage flow), hold on to it for as long as you can (maximize retention time), and if you must let it go, then try to get it back (maximize cycling)" (Fath et al. 2001).

This summary in lay words epitomizes the problem inherent in the search for ecosystem-level maximum principles. In this maxim, the ecosystem is addressed as a person, i.e., as an independent agent that has a goal on its own and whose traits can be selected by natural selection to meet this goal. The similarity to the R^* rule derived from competition theory (chapter 2) is also striking. The R^* rule states that the species with the lowest equilibrium resource requirement wins the competition. But a species' R^* is an aggregate parameter that is determined by its ability to capture resources and to keep them through reduced mortality and other losses. Minimizing R^* is obtained by maximizing per capita resource consumption and resource retention time. Thus, the above maxim is but a reformulation of the R^* rule applied to an ecosystem instead of an individual organism. Clearly, in both form and content, this maxim treats ecosystems as superorganisms.

The superorganism concept has been debated enough in the evolutionary and ecological literature that I do not need to repeat this debate here. The superorganism concept would provide an appropriate representation of reality if ecosystems were fully integrated units subject to pure group selection (Wilson and Sober 1989). I have provided a theoretical example of a superorganismal ecosystem with the plant–decomposer model in a heterogeneous environment above. In that case, the ecosystem behaves as a superorganism because there is a localized, sustained association between individual partners. I analyzed this scenario as a limiting case of pure group selection to explore the potential consequences of group selection on ecosystem properties. But no ecosystem in nature is so fully integrated and localized as to bypass any influence of individual selection. We might expect a combination of individual and group selection to operate under natural conditions, with individual selection often prevailing because many ecological interactions are not strongly localized.

But as soon as individual selection is present, *genuine optimization of ecosystem properties cannot be achieved* because of inevitable *conflicts between levels of selection*. This problem is clearly illustrated in figure 8.4 regarding evolution of plant tolerance to grazing: the optimal level of plant tolerance that yields grazing optimization can never be reached through evolution except under the most extreme scenario of pure group selection. Either individual selection prevails and plant defense is selected for, or there is a combination of individual and group selection and a lower than optimal level of plant tolerance evolves. This is a fairly robust conclusion, which applies to all properties at all levels of selection. Unless the level of selection considered consists of tightly integrated entities (such as individual organisms typically are), selection at lower levels plays a disruptive role because it promotes selfishness at these lower levels, thus yielding nonoptimal behavior at the higher level. As a consequence, properties that are optimal at the ecosystem level are unlikely to evolve by natural selection. Evolved ecosystem properties are likely to express a compromise between different levels of selection when they are not solely driven by individual selection.

Of course, individual selection itself can indirectly lead to "near-optimal" behavior at the ecosystem level. For instance, in the plant–decomposer model in a homogeneous environment examined earlier, which involves only individual selection within material cycles, a number of ecosystem properties increase during succession and evolution, just as they would in the presence of group selection. In this case, the reason is that individual selection and group selection are largely aligned, even though they would likely lead to different optimal configurations if plausible ecological trade-offs were taken into account. As another example, in the model for the evolution of plant defense with a trade-off between antiherbivore defense and nutrient uptake, a certain level of tolerance to grazing evolves, and the evolutionary optimal level that results from individual selection may approach the level that yields grazing optimization. In this case, individual selection would lead to an outcome that is accidentally close to the outcome that is optimal at the ecosystem level. Thus, while genuine optimization of ecosystem properties is extremely unlikely, near optimization is conceivable, either when group selection is strong or when ecological trade-offs happen to make individual selection and group selection converge on roughly similar outcomes. But there is no reason to believe that near optimization of ecosystem properties will generally be achieved, nor is there, for that matter, any objective criterion to measure the extent to which it is achieved.

In conclusion, evolutionary theory is not compatible with genuine maximum principles at the ecosystem level. It is, however, compatible with the fact that ecosystems may approach maximal functioning to some undefined degree. This is perhaps where the crux of the problem lies. Supporters of maximum principles in ecosystem ecology usually do not explicitly distinguish between genuine maximization and mere tendency to increase. And even those who do (e.g., Fath et al. 2001) do not draw all the implications from this distinction. Many of the trends in ecosystem properties identified by E. P. Odum (1969) during the course of succession, such as increased production, increased biomass, decreased production/biomass ratio, and increased cycling efficiency, have an undeniable empirical and theoretical basis (chapter 6). But these trends are not sufficient evidence for maximization or minimization of ecosystem properties. Genuine maximization of an ecosystem property requires not only that this property increases but also that it cannot increase any further. I know of no empirical evidence supporting the latter statement.

I have shown in chapter 4 that trophic interactions in food webs tend to bring each trophic level toward medium, not maximal, functioning. This analysis, however, did not include indirect effects of trophic interactions, in particular, their effects on nutrient cycling, which, as we have seen in later chapters, can lead to enhancement of ecosystem processes and even evolution of indirect mutualism. But even when these positive indirect effects are absent, a successional trend toward increased production and biomass is likely. Consider, for example, the strongly constrained assembly process of the specialist plant–herbivore food web analyzed in figure 4.6. It is reasonable to expect that the community will not reach equilibrium at every stage of the assembly process, which will considerably attenuate the fluctuations in the amount of available inorganic nutrient. In fact, if herbivores invade relatively quickly after their host plants, the successional sequence will roughly follow the right-hand side of figure 4.6 and generate a more or less gradual decrease in the amount of available inorganic nutrient accompanied by corresponding increases in total biomass and production. Yet, an ecosystem that lacks herbivores would be more productive than the diverse ecosystem that results from this assembly process. Thus, an increase in ecosystem properties during succession is not equivalent to maximization of these properties.

Statistical principles from thermodynamics and statistical mechanics are often used to support maximum principles in ecology, which raises another important issue, that of the epistemological status of these physical laws and their applicability to ecology. In particular, strong theoretical and empirical support has now accumulated for the maximum entropy production principle in nonequilibrium statistical mechanics (Dewar 2003a). But it is important to understand the scope and limits of this principle. Maximum entropy production occurs in nonequilibrium thermodynamic systems for the same reason that maximum entropy occurs in equilibrium thermodynamic systems—because it is the most probable state, i.e., the state that can be realized by more microscopic pathways than any other (Dewar 2003b). As a consequence, for the maximum entropy or maximum entropy production principles to be of any predictive value, the probability distribution of microscopic states must be sufficiently concentrated such that one state is much more probable than others. This occurs in statistical mechanics because of the huge number of particles involved (typically on the order of Avogadro's number, i.e., 6×10^{23} particles per mole), which creates a clear-cut separation between the microscopic and macroscopic scales and a strongly peaked probability distribution of microscopic states. Such a separation of scales does not occur in ecology, at least at the scale of a local ecosystem. Individual organisms—the particles of ecology— belong to myriad different species that occupy different niches, and their numbers are much smaller than those of particles in statistical mechanics. O'Neill et al. (1986) christened ecosystems "medium-number systems" to account for this property. As a consequence, blind applications of entropy maximization algorithms in ecology can lead to serious flaws and misinterpretations (Haegeman and Loreau 2008, 2009).

Although it originated in statistical mechanics (Jaynes 1957), entropy maximization has developed into a general statistical inference technique rooted in information theory, with applications in a wide range of disciplines (Jaynes 2003). That is why the scope and interpretation of this approach have been debated. There are two rationales behind entropy maximization, which are but expressions of the two interpretations of probability theory (Haegeman and Loreau 2008, 2009). The first rationale is that entropy maximization predicts the macroscopic state of a system that can be realized by the largest number of microscopic states under some macroscopic constraints. This is the classical interpretation originating from statistical mechanics, which is rooted in frequency-based probability theory. On this interpretation, separation between microscopic and macroscopic scales is a key feature that gives entropy maximization its predictive power. The second rationale is rooted in information theory and Bayesian probability theory. In information theory, entropy measures uncertainty, or the amount of missing information. Entropy maximization then yields the macroscopic state that maximizes our uncertainty regarding all we do not know about a problem after having accounted for known macroscopic constraints. Separation between microscopic and macroscopic scales is not critical in this interpretation. The two interpretations, however, are entirely complementary and provide different perspectives on the same problem. When the macroscopic and microscopic scales strongly overlap, entropy maximization still provides the least biased statistical inference, but its predictive power is very low and it leads to largely trivial applications that do not bring any new insights compared with the initial data (Haegeman and Loreau 2008, 2009).

Perhaps ecology has simply not looked at the appropriate scale and issues to apply the principles and tools of statistical mechanics. Macroecology is a research program that seeks to expand the scale of ecological investigations in the hope of uncovering statistical laws at large spatial scales (Brown 1995; Maurer 1999). Approaches borrowed from statistical mechanics are likely to be more relevant at these scales, although convincing applications are still rare (Harte et al. 2008). Neutral theory is another area that lends itself to application of statistical mechanics because it makes the assumption that all species are equivalent, thereby effectively reducing the actual complexity of natural communities to an ensemble of idealized identical "particles." Nevertheless, predicting communitywide patterns such as species abundance distributions requires more microscopic details than does the prediction of macroscopic properties in thermodynamics. Therefore, the microscopic and macroscopic scales are less separated, which may limit the relevance of statistical mechanical approaches in this case.

CONCLUSION

Bridging the historical gap between ecosystem ecology and evolutionary biology requires a dialogue between the two disciplines based not only on mutual understanding but also on scientific rigor. In this chapter, I have attempted to develop the bases of such an open but rigorous dialogue. I have shown that ecosystems and their properties can evolve as the result of at least three nonexclusive processes that require increasingly stringent conditions, i.e., individual-level selection of species that indirectly affect ecosystem functioning, evolutionary feedback between organisms and ecosystem properties, and ecosystem-level selection. I have discussed several theoretical examples of these processes while examining some fascinating aspects of the evolution of plant–decomposer and plant–herbivore interactions within an ecosystem context and the evolutionary emergence of entire food webs.

One recurrent theme throughout this chapter is that conflicts between levels of selection are likely to be common and preclude any genuine maximization or optimization of ecosystem properties through evolution. The fact that evolution and community assembly do not lead to optimal ecosystem configurations, however, does not contradict the existence of directional trends during succession. In fact, recognizing the difference between genuine maximization and directional successional trends might be one of the keys to reconciling the perspectives of ecosystem ecologists and evolutionary biologists.

Another of these keys might be a better understanding of the value and limitations of approaches borrowed from statistical mechanics in ecology.

These approaches would be extremely useful to help make ecology a more predictable science, but they have often been applied too uncritically so far. Convincing applications at relevant scales are still to come. Perhaps then will we be able to identify and understand robust "maximum principles" in ecology. Developing a sound statistical mechanical approach might be one of the most important future challenges for ecology.

Postface: Toward an Integrated, Predictive Ecology

The human species is arguably at a turning point in its historical development. In a few millennia, humans have risen from the state of sparse populations of gatherers-hunters with minor impacts on their environment to that of a global collective force that is reshaping the face of Earth. The fate of our planet hinges to a significant extent on how humankind will handle its new status of global dominant species and adapt its behavior and society accordingly during this century. The global human population and economy are still growing nearly exponentially today. One of the most striking and yet poorly appreciated properties of exponential growth is that a population can grow in absolute terms as much in a single generation as during its entire previous history. By contrast, planet Earth has a constant size. Since the fraction of land transformed or degraded by humanity was estimated to lie somewhere between 39 percent and 50 percent more than a decade ago (Vitousek et al. 1997), humankind could easily destroy all remaining "natural" or unmanaged ecosystems during this century, with devastating consequences for biodiversity, global biogeochemical cycles, the global climate, and, indirectly, humans themselves. Whether we like it or not, a global environmental crisis lies before us with certainty—in fact, it has already started. What is less certain, and gives a glimmer of hope, is how humanity will face this crisis. Any crisis is an opportunity to transform oneself. If humankind manages to radically change its social relationships and its relationship to nature, it can establish a new balance within nature that meets its own needs while preserving the resources and beauty of the natural environment in which it has grown (Loreau 2010).

The global environmental crisis we are entering has several components. Chemical pollution and now climate change have attracted a lot of attention from both scientists and the general public, in part because they are relatively easily measured and monitored. But perhaps an even greater challenge is the future of life and its diversity on our planet (Wilson 1992,

2002; Loreau 2010). Humankind is destroying biological diversity and natural ecosystems at rates unprecedented since the mass extinction episode that marked the end of dinosaurs 65 million years ago. The causes of this biodiversity loss and the forms it takes are manifold, but they can all be traced back to the rapid expansion of the global human population and the concomitant expansion of global production, consumption, and trade. Its consequences are potentially considerable and still surprisingly underestimated. They include a decreased diversity and increasing vulnerability of our food resources, the loss of an irreplaceable natural heritage with considerable spiritual, aesthetic, and recreational values, and the degradation of a wide range of ecosystem services that we take for granted, such as crop pollination, water supply and purification, regulation of air quality and climate, and regulation of soil erosion and fertility. Their general effect is a decline in the quality of life. By and large humanity still views progress as the unlimited expansion of the human enterprise, as though nature were infinitely plentiful and technology infinitely powerful. But this traditional view of progress is no longer tenable today. If progress has any meaning, it should be focused on the quality of life and take into account the long-term consequences of our actions. Therefore, massive biodiversity loss is a serious threat to progress, and its multiple consequences need to be understood thoroughly.

As the science of the relationships between organisms and their environment, ecology is set to gain increasing importance during this century. But in order to help human societies face growing environmental challenges, ecology will have to deliver relevant scientific knowledge on how natural ecosystems function, how changes in biodiversity affect them, how both biodiversity and ecosystems are inextricably linked to human well-being, and how humankind can interact with them in a sustainable way. And this requires that ecology transform itself into a more integrated, predictive science. The traditional slicing of ecology based on organizational levels into autecology, population ecology, community ecology, and ecosystem ecology has some value because different organizational levels obey partly different sets of constraints and thus partly different laws. But it also has strong limitations because it tends to perpetuate arbitrary divisions and hamper the emergence of unifying perspectives. I have shown in this book that populations, communities, and ecosystems are but different perspectives on the same material reality and that merging these perspectives and those of evolutionary biology provides important new insights into the dynamics and functioning of ecological systems. Not only does this synthesis offer exciting new avenues of basic research but it also illuminates links between

biodiversity, species interactions, and ecosystem functioning that are critical to predict future impacts of anthropogenic environmental changes on
the Earth system and on human societies.

FOUNDATIONS FOR A NEW ECOLOGICAL SYNTHESIS

A new ecological synthesis can emerge only if the conceptual and theoretical foundations of the various ecological subdisciplines can be made compatible with each other, such that the fundamental insights and results
from each subdiscipline can be translated into the language of the others
and hence be incorporated into their theoretical arsenal. I have shown that
the principles of population dynamics and ecosystem functioning can be
made compatible through their joint dependence on the mass and energy
budgets of individual organisms. Consistent models that merge the population and ecosystem perspectives can then be obtained by coupling the
formalism of compartmental models borrowed from ecosystem ecology
and the versatility of nonlinear functions that determine mass and energy
flows borrowed from population and community ecology (chapter 1).

I have used this approach throughout the book to develop new perspectives on the dynamics and functioning of ecosystems that transcend the
boundaries of traditional ecological subdisciplines. To do so, I have followed a logical progression from simpler to more complex systems, and
from smaller to larger spatial and temporal scales. Two conceptual steps
are involved in the transition from populations to ecosystems. The first step
consists in linking the properties of individual populations to the aggregate
properties of communities made up of species with similar ecologies, i.e.,
species that belong to the same functional group or trophic level. The second step consists in linking the properties of the various functional groups
or trophic levels to the functioning of the ecosystem as a whole. Both steps
are key to understanding the relationships between biodiversity and ecosystem functioning. The first step allows examining the functional consequences of horizontal diversity, the second, those of vertical diversity.

Competition theory provides the tools to make the first step of this
transition since it addresses the interactions that emerge between species
that share common resources. Competition theory, however, has been
used traditionally to examine the conditions that allow species to coexist
and species diversity to be maintained. Here I have revisited it to examine
how species coexistence affects the functional properties of communities.
My broad overview of mechanisms of species coexistence concludes that

species diversity should have some fairly general functional consequences in competitive communities, though none is universal (chapter 2). In particular, competition theory predicts that diverse communities should generally show more efficient resource exploitation than do single species, and hence overyielding. By contrast, transgressive overyielding is expected to occur only under specific coexistence mechanisms and requires large enough degrees of niche differentiation. These predictions have been largely confirmed by recent theoretical and experimental work on the impacts of biodiversity on ecosystem functioning. This work shows that species diversity does have the potential to affect ecosystem processes to a measurable extent, and that some form of functional complementarity between species driven by niche differentiation or facilitation is responsible for the effects of species diversity (chapter 3). Yet, despite these effects, transgressive overyielding appears to be relatively uncommon in recent experiments. Thus, we now have a consistent body of theory and experimental work that provides the fundamental bases to predict the effects of biodiversity changes on ecosystem functioning and hence potentially on the delivery of ecosystem services to human societies.

Linking stability properties at the population level and at the aggregate community level is also providing a radically new perspective on the old debate about the relationship between the complexity and stability of ecological systems. New theory shows that fluctuations in the abundance of individual populations can yield stable functioning of the community as a whole. Differences among species in their responses to environmental fluctuations are the main mechanism that generates a stabilizing effect of species diversity on aggregate community properties (chapter 5). This theory is a major step forward because it helps to reconcile previous theoretical results on the destabilizing effect of species diversity, which focused implicitly on population-level stability, and empirical observations that diverse ecosystems are often stable. It also provides significant new insights into the long-term consequences of biodiversity changes. Biodiversity loss is likely not only to alter the magnitude of ecosystem processes but also to increase their variability, and, accordingly, to decrease the reliability of the delivery of ecosystem services.

Linking functional groups or trophic levels and whole ecosystems is a second necessary step in the transition from populations to ecosystems. I have taken this step in two parts. First I have examined the properties of food webs or interaction webs connected by trophic and nontrophic interactions between species that belong to multiple trophic levels. A striking conclusion that emerges from this theoretical work is that the relationships between biodiversity and ecosystem functioning are potentially more complex when multiple trophic levels interact. In particular, in contrast to horizontal diversity within trophic levels, vertical diversity between trophic levels does not appear to maximize any ecosystem property. Instead, it makes whole-trophic-level properties converge on intermediate values through damped oscillations as food-chain length increases. Overexploitation is another factor that can oppose the positive effects of horizontal diversity and deteriorate the functioning of diverse, strongly interacting ecosystems. There is no ground for the commonly held belief that complex ecosystems should function optimally (chapter 4). Yet, despite these complications, it is remarkable that horizontal species diversity can stabilize ecosystem properties in food webs just as in simple competitive communities, provided consumers are sufficiently specialized or bear a cost for being generalists (chapter 5). Interactions between multiple trophic levels are a major source of potential surprises and uncertainty in our predictions of the effects of biodiversity loss on ecosystem services as small biodiversity changes can trigger abrupt changes in the structure and functioning of complex ecosystems.

I have then examined the specific role of material cycling in ecosystem functioning. Material cycling is a powerful whole-ecosystem process that binds all ecosystem components together and transmits positive indirect effects to all of them, thereby contributing to mitigate, or even reverse, the negative effects of trophic interactions. Through its effects on the input–output balance of limiting nutrients at the ecosystem level, it yields simple, general laws that govern equilibrium ecosystem properties. One of its most important properties is that it couples the productions of the various living compartments, which tends to generate indirect mutualism between ecosystem components. Combined with the dynamics of interspecific competition, it also provides a mechanistic explanation for general trends in ecosystem properties during succession, including increased biomass, increased production, decreased productivity per unit biomass, and increased nutrient cycling efficiency (chapter 6). Thus, material cycling is an ecosystem process that potentially impinges on the ecological and evolutionary dynamics of all the species that make up the ecosystem. Understanding its properties is critical to predict the long-term ecological consequences of changes in both biodiversity and environmental factors, as any biotic or abiotic change that affects inputs and outputs of limiting nutrients may have considerable effects on ecosystem processes and services in the long run.

The final two chapters extend the scope of my exploration of the interplay between community dynamics and ecosystem functioning to larger spatial and temporal scales. Metacommunity and metaecosystem theory offers a promising framework to explore the emergent properties that arise from spatial coupling of communities and ecosystems, thus providing new insights into biodiversity and ecosystem functioning from local to global scales. Metacommunity or metaecosystem connectivity, as determined by the spatial arrangement of component ecosystems and the movements of organisms, energy, and inorganic substances across these ecosystems, exerts strong constraints on biodiversity, ecosystem functioning, and the relationship between them at both local and regional scales. It also drives many of the community and ecosystem properties that are traditionally studied at separate scales without consideration of these critical connections among scales (chapter 7). Therefore, this new body of theory at the interface between community and ecosystem ecology is likely to enhance our ability to predict the consequences of land-use changes on biodiversity and the provision of ecosystem services to human societies.

Last, I have paved the way for a much needed synthesis of ecosystem ecology and evolutionary biology. Bridging the historical gap between the two disciplines requires a dialogue based not only on mutual understanding but also on scientific rigor. I have sought to establish the bases of such a dialogue by identifying the various processes by which natural selection can lead to the evolution of ecosystems and ecosystem properties, and by discussing several theoretical examples of these processes bearing on the evolution of plant–decomposer and plant–herbivore interactions within an ecosystem context and on the evolutionary emergence of entire food webs (chapter 8). One robust conclusion that emerges from these considerations is that conflicts between levels of selection are common and preclude genuine maximization or optimization of ecosystem properties through evolution. Building an evolutionary ecosystem ecology may be one of the most important challenges of ecology today. The long-term consequences of environmental changes cannot be understood without proper consideration of evolutionary dynamics and its effects on ecosystem functioning.

Each aspect of the transition from populations to ecosystems that I have explored in this book is bringing new results and principles that improve our ability to understand and predict the functioning of ecosystems and their responses to biodiversity loss and other environmental changes. These new results and principles show the value and power of a synthetic approach to ecology that bridges the gaps between its various subdisciplines. They do not, however, constitute an accomplished ecological synthesis. Unfortunately, there are still too many gaps in our understanding of complex ecological systems to reach this goal today.

FUTURE CHALLENGES

No theoretical unification can pretend to encompass the whole of reality and resolve all issues, and this book is no exception. I have already alluded to some of the future challenges that arise naturally from the material covered in the previous chapters. But the trend toward unification of ecology should not stop at the borders of population and ecosystem ecology. Perhaps the greatest future challenges for both theoretical and empirical ecology will lie outside the traditional "boxes," at the interface with other sciences. I can envision at least three areas in which theoretical advances would be particularly valuable to help transforming ecology into a more integrated, predictive science, and which I have not tackled in this book.

First is the *explicit incorporation in ecological theory of physical and chemical properties of both organisms and their environment*. Theoretical ecology developed asymmetrically during the second half of the 20th century, with a much greater emphasis on population and communities than on ecosystems. In this book I have shown the tremendous potential of merging the demographic approach of population and community ecology and the functional approach of ecosystem ecology. But I have deliberately chosen to explore a limited set of basic constraints and properties that arise from the transfer and cycling of a single nutrient within ecosystems. In reality, ecological systems are constrained by a large set of physico-chemical factors that act as either resources or environmental conditions for organisms. Basal inorganic resources that limit primary production, and hence indirectly the whole food web, include nitrogen, phosphorus, carbon, water, and energy, the availability of which can be partly decoupled in space and time. Elementary physicochemical properties of the ambient medium, such as temperature, humidity, density, light absorption, and diffusion of minerals, also determine the supply of these resources to organisms and the amounts of energy and materials they spend in feeding, surviving, and reproducing. Therefore, a comprehensive theory of ecology and evolution should ideally include all the physical and chemical factors that affect the fitness of organisms, the interactions between species, and the structure and functioning of ecosystems.

Marine ecology has traditionally paid more attention to these factors than has terrestrial ecology because of the stronger influence of physical factors in the marine environment. Ecological stoichiometry (Sterner and Elser 2002) and metabolic theory (Brown et al. 2004) have started to examine a few of these factors from a more general perspective. But we are still far from a complete integration of physicochemical factors in ecological theory. Such integration concerns not only the effects of abiotic factors on organisms but also the effects that organisms in turn have on these factors over ecological and evolutionary time scales. The concepts of ecosystem engineering (Jones et al. 1994) and niche construction (Odling-Smee et al. 2003) seek to capture this active transformational role of organisms on their abiotic environment. The ecological and evolutionary feedbacks between organisms and their environment, however, are still poorly understood (Kylafis and Loreau 2008). Yet, they are critical to understand the evolutionary history of the Earth system and the impacts of humans on this system.

Second, *the development of new approaches that link genes, traits, and functions* is another promising direction for ecological theory that would considerably strengthen its integrative and predictive potential. The species concept occupies a central role in current ecological and evolutionary thinking for good reasons. But the species concept also has limitations when it comes to describing biodiversity, understanding ecosystem functioning, and predicting the effects of future environmental changes on the delivery of ecosystem services to humans. Microbial diversity, for instance, is increasingly studied using molecular techniques that do not call for the identification of species but that provide direct information on genetic or functional diversity. With the rapid development of genomics and proteomics, biodiversity data might soon be available in the form of genes or physiological functions, even for large organisms. Since the functional role of biodiversity in ecosystems is fundamentally governed by functional trait diversity (chapters 3 and 5), not species diversity per se, molecular data might eventually prove handier than species data to predict the impacts of biodiversity changes on ecosystem services. The challenge for theoretical ecology will then be to establish a new coherent framework that links genes, phenotypic traits, and ecological functions, thus downplaying the species level in the description of ecological systems. Although still in their infancy, trait-based approaches are receiving increasing attention in both community ecology and ecosystem ecology (McGill et al. 2006a; Savage et al. 2007; Suding et al. 2008) and constitute a promising step in that direction.

A third avenue for the development of ecological theory that will be increasingly needed in the future is the *integration of ecology and economics.* Ecological and economic systems share a number of properties—in particular, they are complex adaptive systems in which a large number of agents interact both competitively and cooperatively in the consumption and production of a finite amount of products and resources. Although several concepts have permeated historically through the boundary of the two disciplines and there is even a growing subdiscipline of economics—ecological economics—that seeks to incorporate ecological constraints into economic thinking, ecology and economics are still remarkably distant overall in their objectives and approaches. Given the fast-growing clash and interpenetration of humans and nature that generates the current global environmental crisis, a much stronger integration of ecology and economics will be necessary if we are to resolve this crisis. A fast-growing current trend is to translate ecosystem services into economic values, which contributes to raising societal awareness of our dependence on ecosystems and accounting for some of the detrimental environmental consequences of our actions (Heal et al. 2005; MEA 2005). But incorporating ecological constraints into an autonomous economic system will not be enough in the long run. Integrating ecology and economics means, more fundamentally, reinserting humans where they belong, i.e., within nature. Accordingly, we need to build a theory that allows us to conceptualize, formalize, and quantify this insertion of humans in nature, of economy in ecology.

It is my firm belief that ecological theory should be both a guide for basic research and a guide for action. Developing a new ecological synthesis that moves beyond traditional disciplinary boundaries to uncover general principles and broader perspectives is not only a fruitful way to explore and illuminate exciting new scientific issues. It is also a necessity if we are to make ecology a powerful tool for surmounting the looming global environmental crisis. Only a synthetic view can take into account the multiple interactions and feedbacks between organizational levels and spatial and temporal scales that are the hallmark of ecological systems. Therefore, only a synthetic view can allow us to understand and predict the ecological and social consequences of the many changes we are currently bringing about in our local and global environment.

References

Aarssen, L. W. 1997. High productivity in grassland ecosystems: effected by species diversity or productive species? Oikos 80:183–184.

Abbadie, L., A. Mariotti, and J.-C. Menaut. 1992. Independence of savanna grasses from soil organic matter for their nitrogen supply. Ecology 73:608–613.

Abrams, P. A. 1976. Niche overlap and environmental variability. Mathematical Biosciences 28:357–372.

———. 1983. The theory of limiting similarity. Annual Review of Ecology and Systematics 14:359–376.

———. 1988. How should resources be counted? Theoretical Population Biology 33:226–242.

———. 1993. Effect of increased productivity on the abundances of trophic levels. American Naturalist 141:351–371.

———. 1995. Implications of dynamically variable traits for identifying, classifying, and measuring direct and indirect effects in ecological communities. American Naturalist 146:112–134.

———. 1998. High competition with low similarity and low competition with high similarity: exploitative and apparent competition in consumer-resource systems. American Naturalist 152:114–128.

Abrams, P. A., and L. R. Ginzburg. 2000. The nature of predation: prey dependent, ratio dependent or neither? Tree 15:337–341.

Abrams, P. A., C. Rueffler, and R. Dinnage. 2008. Competition-similarity relationships and the nonlinearity of competitive effects in consumer-resource systems. American Naturalist 172:463–474.

Adler, F. R., and J. Mosquera. 2000. Is space necessary? Interference competition and limits to biodiversity. Ecology 81:3226–3232.

Agren, G. I., and E. Bosatta. 1996. Theoretical ecosystem ecology: understanding element cycles. New York, Cambridge University Press.

Amarasekare, P. 2003. Competitive coexistence in spatially structured environments: a synthesis. Ecology Letters 6:1109–1122.

Anstett, M. C., M. Hossaert-McKey, and F. Kjellberg. 1997. Figs and fig pollinators: evolutionary conflicts in a coevolved mutualism. Trends in Ecology and Evolution 12:94–99.

Arditi, R., and L. R. Ginzburg. 1989. Coupling in predator-prey dynamics: ratio-dependence. Journal of Theoretical Biology 139:311–326.

Arditi, R., J. Michalski, and A. H. Hirzel. 2005. Rheagogies: modelling non-trophic effects in food webs. Ecological Complexity 2:249–258.

Armstrong, R. A., and R. MacGehee. 1980. Competitive exclusion. American Naturalist 115:151–170.

Armsworth, P. R., and J. E. Roughgarden. 2003. The economic value of ecological stability. Proceedings of the National Academy of Sciences of the USA 100: 7147–7151.

Arthur, W. 1987. The niche in competition and evolution. Chichester, John Wiley & Sons.

Bascompte, J., P. Jordano, C. J. Melian, and J. M. Olesen. 2003. The nested assembly of plan-animal mutualistic networks. Proceedings of the National Academy of Science of the USA 100:9383–9387.

Baumgärtner, S. 2007. The insurance value of biodiversity in the provision of ecosystem services. Natural Resource Modeling 20:87–127.

Begon, M., J. L. Harper, and C. R. Townsend. 1996. Ecology: individuals, populations and communities. III. London, Blackwell Scientific Publications.

Bell, G. 2007. The evolution of trophic structure. Heredity 99:494–505.

Belsky, A. J. 1986. Does herbivory benefit plants? A review of the evidence. American Naturalist 127:870–892.

Belsky, A. J., W. P. Carson, C. L. Jensen, and G. A. Fox. 1993. Overcompensation by plants: herbivore optimization or red herring? Evolutionary Ecology 7:109–121.

Bolker, B. M., and S. W. Pacala. 1999. Spatial moment equations for plant competition: understanding spatial strategies and the advantages of short dispersal. American Naturalist 153:575–602.

Bond, E. M., and J. M. Chase. 2002. Biodiversity and ecosystem functioning at local and regional spatial scales. Ecology Letters 5:467–470.

Bronstein, J. L. 1994. Our current understanding of mutualism. Quarterly Review of Biology 69:31–51.

Brose, U., R. J. Williams, and N. D. Martinez. 2006. Allometric scaling enhances stability in complex food webs. Ecology Letters 9:1228–1236.

Brown, J. H. 1995. Macroecology. Chicago, University of Chicago Press.

Brown, J. H., J. F. Gillooly, A. P. Allen, V. M. Savage, and G. B. West. 2004. Toward a metabolic theory of ecology. Ecology 85:1771–1789.

Bruno, J. F., J. J. Stachowicz, and M. D. Bertness. 2003. Inclusion of facilitation into ecological theory. Trends in Ecology & Evolution 18:119–125.

Cadotte, M. W. 2006. Dispersal and species diversity: a meta-analysis. American Naturalist 167:913–924.

Caldarelli, G., P. G. Higgs, and A. J. McKane. 1998. Modelling coevolution in multispecies communities. Journal of Theoretical Biology 193:345–358.

Callaway, R. M., and L. R. Walker. 1997. Competition and facilitation: a synthetic approach to interactions in plant communities. Ecology 78:1958–1965.

Cardinale, B. J., M. A. Palmer, and J. P. Collins. 2002. Species diversity enhances ecosystem functioning through interspecific facilitation. Nature 415:426–429.

Cardinale, B. J., D. S. Srivastava, J. E. Duffy, J. P. Wright, A. L. Downing, M. Sankaran, and C. Jouseau. 2006. Effects of biodiversity on the functioning of trophic groups and ecosystems. Nature 443:989–992.

Cardinale, B. J., J. P. Wright, M. W. Cadotte, I. T. Carroll, A. Hector, D. S. Srivastava, M. Loreau and J. J. Weis. 2007. Impacts of plant diversity on biomass production increase through time because of species complementarity. Proceedings of the National Academy of Sciences of the USA 104:18123–18128.

Carpenter, S. R., J. F. Kitchell, and J. R. Hodgson. 1985. Cascading trophic interactions and lake productivity. BioScience 35:634–639.

Case, T. J. 2000. An illustrated guide to theoretical ecology. Oxford, Oxford University Press.

Case, T. J., R. D. Holt, M. A. McPeek, and T. H. Keitt. 2005. The community context of species' borders: ecological and evolutionary perspectives. Oikos 108:28–46.

Casula, P., A. Wilby, and M. B. Thomas. Understanding biodiversity effects on prey in multi-enemy systems. Ecology Letters 9:995–1004.

Caswell, H. 1989. Matrix population models. Sunderland, Massachusetts, Sinauer Associates, Inc.

Cattin, M. F., L. F. Bersier, C. Banasek-Richter, R. Baltensperger, and J. P. Gabriel. 2004. Phylogenetic constraints and adaptation explain food-web structure. Nature 427:835–839.

Chase, J. M., and M. A. Leibold. 2002. Spatial scale dictates the productivity-biodiversity relationship. Nature 416:427–430.

———. 2003. Ecological niches: linking classical and contemporary approaches. Chicago, University of Chicago Press.

Chave, J. 2004. Neutral theory and community ecology. Ecology Letters 7:241–253.

Cherif, M., and M. Loreau. 2007. Stoichiometric constraints on resource use, competitive interactions, and elemental cycling in microbial decomposers. American Naturalist 169:709–724.

Chesson, P. 1994. Multispecies competition in variable environments. Theoretical Population Biology 45:227–276.

———. 2000a. General theory of competitive coexistence in spatially varying environments. Theoretical Population Biology 58:211–237.

———. 2000b. Mechanisms of maintenance of species diversity. Annual Review of Ecology and Systematics 31:343–366.

Chesson, P., and N. Huntly. 1997. The roles of harsh and fluctuating conditions in the dynamics of ecological communities. American Naturalist 150:519–553.

Chesson, P., S. Pacala, and C. Neuhauser. 2001. Environmental niches and ecosystem functioning. Pages 213–262 in A. Kinzig, S. Pacala, and D. Tilman, eds. The functional consequences of biodiversity: empirical progress and theoretical extensions. Princeton, New Jersey, Princeton University Press.

Christensen, K., S. A. di Collobiano, A. R. Hall, and H. J. Jensen. 2002. Tangled nature: a model of evolutionary ecology. Journal of Theoretical Biology 216:73–84.

Cohen, J. E., F. Briand, and C. M. Newman. 1990. Community food webs: data and theory Berlin, Springer-Verlag.

Connell, J. H. 1978. Diversity in tropical rain forests and coral reefs. Science 199:1302–1310.

———. 1983. On the prevalence and relative importance of interspecific competition: evidence from field experiments. American Naturalist 122:661–696.

Connell, J. H., and R. O. Slatyer. 1977. Mechanisms of succession in natural communities and their role in community stability and organization. American Naturalist 111:1119–1144.

Cornell, H. V. 1993. Unsaturated patterns in species assemblages: the role of regional proccesses in setting local species richness. Pages 243–252 *in* R. E. Ricklefs and D. Schluter, eds. Species diversity in ecological communities: historical and geographical perspectives. Chicago, University of Chicago Press.

Cornell, H. V., and J. H. Lawton. 1992. Species interactions, local and regional processes, and limits to the richness of ecological communities: a theoretical perspective. Journal of Animal Ecology 61:1–12.

Crowley, P. H. 1992. Resampling methods for computation-intensive data analysis in ecology and evolution. Annual Review of Ecology and Systematics 23:405–447.

Danovaro, R., C. Gambi, A. Dell'Anno, C. Corinaldesi, S. Fraschetti, A. Vanreusel, M. Vincx and A. J. Gooday. 2008. Exponential decline of deep-sea ecosystem functioning linked to benthic biodiversity loss. Current Biology 18:1–8.

Daufresne, T., and M. Loreau. 2001. Ecological stoichiometry, primary producer-decomposer interactions, and ecosystem persistence. Ecology 82:3069–3082.

Dawkins, R. 1982. The extended phenotype: the gene as the unit of selection. Oxford, Freeman.

de Mazancourt, C., E. Johnson, and T. G. Barraclough. 2008. Biodiversity inhibits species' evolutionary responses to changing environments. Ecology Letters 11:380–388.

de Mazancourt, C., and M. Loreau. 2000a. Effect of herbivory and plant species replacement on primary production. American Naturalist 155:735–754.

———. 2000b. Grazing optimization, nutrient cycling and spatial heterogeneity of plant-herbivore interactions: should a palatable plant evolve? Evolution 54:81–92.

de Mazancourt, C., M. Loreau, and L. Abbadie. 1998. Grazing optimization and nutrient cycling: when do herbivores enhance plant production? Ecology 79:2242–2252.

———. 1999. Grazing optimization and nutrient cycling: potential impact of large herbivores in a savanna system. Ecological Applications 9:784–797.

de Mazancourt, C., M. Loreau, and U. Dieckmann. 2001. Can the evolution of plant defense lead to plant-herbivore mutualism? American Naturalist 158:109–123.

———. 2005. Understanding mutualism when there is adaptation to the partner. Journal of Ecology 93:305–314.

De Roos, A. M., L. Persson, and E. McCauley. 2003. The influence of size-dependent life-history traits on the structure and dynamics of populations and communities. Ecology Letters 6:473–487.

de Ruiter, P. C., A.-M. Neutel, and J. C. Moore. 1995. Energetics, patterns of interaction strengths, and stability in real ecosystems. Science 269:1257–1260.

De Wit, C. T. 1960. On competition. Verslag Landbouwkundig Onderzoek. Wageningen, Netherlands.

De Wit, C. T., and J. P. van der Bergh. 1965. Competition between herbage plants. Netherlands Journal of Agricultural Science 13:212–221.

DeAngelis, D. L. 1975. Stability and connectance in food web models. Ecology 56:238–243.

———. 1992. Dynamics of nutrient cycling and food webs: population and community biology. London, Chapman & Hall.

DeAngelis, D. L., R. A. Goldstein, and R. V. O'Neill. 1975. A model for trophic interaction. Ecology 56:881–892.

DeAngelis, D. L., and J. Gross. 1992. Individual-based models and approaches in ecology: populations, communities, and ecosystems. New York, Chapman & Hall.

Dewar, R. 2003a. Information theory explanation of the fluctuation theorem, maximum entropy production and self-organized criticality in non-equilibrium stationary states. Journal of Physics A 36:631–641.

Dewar, R. C. 2003b. Maximum entropy production and non-equilibrium statistical mechanics. Pages 41–55 in A. Kleidon and R. D. Lorenz, eds. Non-equilibrium thermodynamics and the production of entropy: life, Earth, and beyond. Berlin, Springer.

Diaz, S., and M. Cabido. 2001. Vive la différence: plant functional diversity matters to ecosystem processes. Trends in Ecology and Evolution 16:646–655.

Dieckmann, U., R. Law, and J. A. J. Metz. 2000. The geometry of ecological interactions: simplifying spatial complexity: Cambridge studies in adaptive dynamics. Cambridge, Cambridge University Press.

Dimitrakopoulos, P. G., and B. Schmid. 2004. Biodiversity effects increase linearly with biotope space. Ecology Letters 7:574–583.

Doak, D. F., D. Bigger, E. K. Harding, M. A. Marvier, R. E. O'Malley, and D. Thomson. 1998. The statistical inevitability of stability-diversity relationships in community ecology. American Naturalist 151:264–276.

Downing, K., and P. Zvirinsky. 1999. The simulated evolution of biochemical guilds: reconciling Gaia theory and natural selection. Artificial Life 5:291–318.

Downing, K. L. 2002. The simulated emergence of distributed environmental control in evolving microcosms. Artificial Life 8:123–153.

Drossel, B., P. G. Higgs, and A. J. McKane. 2001. The influence of predator-prey population dynamics on the long-term evolution of food web structure. Journal of Theoretical Biology 208:91–107.

Duffy, J. E., B. J. Cardinale, K. E. France, P. B. McIntyre, E. Thebault, and M. Loreau. 2007. The functional role of biodiversity in ecosystems: incorporating trophic complexity. Ecology Letters 10:522–538.

Duffy, J. E., J. P. Richardson, and E. A. Canuel. 2003. Grazer diversity effects on ecosystem functioning in seagrass beds. Ecology Letters 6:637–645.

Dunbar, M. J. 1960. The evolution of stability in marine environments: natural selection at the level of the ecosystem. American Naturalist 94:129–136.

Economo, E. P., and T. H. Keitt. 2008. Species diversity in neutral metacommunities: a network approach. Ecology Letters 11:52–62.

Elser, J. J., M. E. S. Bracken, E. E. Cleland, D. S. Gruner, W. S. Harpole, H. Hillebrand, J. T. Ngai, E. W. Seabloom, J. B. Shurin, and J. E. Smith. 2007. Global analysis of nitrogen and phosphorus limitation of primary producers in freshwater, marine and terrestrial ecosystems. Ecology Letters 10:1135–1142.

Elton, C. S. 1927. Animal ecology. London, Sidgwick and Jackson.

———. 1958. The ecology of invasions by animals and plants. London, Methuen.

Engen, S., R. Lande, B. E. Saether, and T. Bregnballe. 2005. Estimating the pattern of synchrony in fluctuating populations. Journal of Animal Ecology 74: 601–611.

Ernest, S. K. M., and J. H. Brown. 2001. Homeostasis and compensation: the role of species and resources in ecosystem stability. Ecology 82:2118–2132.

Faith, D. P. 1992. Conservation evaluation and phylogenetic diversity. Biological Conservation 61:1–10.

Fargione, J., D. Tilman, R. Dybzinski, J. Hille Ris Lambers, C. Clark, W. S. Harpole, J. M. H. Knops, P. B. Reich, and M. Loreau. 2007. From selection to complementarity: shifts in the causes of biodiversity-productivity relationships in a long-term biodiversity experiment. Proceedings of the Royal Society B274:871–876.

Fath, B. D., S. E. Jorgensen, B. C. Patten, and M. Straskraba. 2004. Ecosystem growth and development. BioSystems 77:213–228.

Fath, B. D., B. C. Patten, and J. S. Choi. 2001. Complementarity of ecological goal functions. Journal of Theoretical Biology 208:493–506.

Finke, D. L., and R. F. Denno. 2005. Predator diversity and the functioning of ecosystems: the role of intraguild predation in dampening trophic cascades. Ecology Letters 8:1299–1306.

Finn, J. T. 1982. Ecosystem succession, nutrient cycling and output-input ratios. Journal of Theoretical Biology 99:479–489.

Forman, R. T. T. 1995. Land mosaic: the ecology of landscapes and regions. Cambridge, Cambridge University Press.

Fox, J. W. 2005. Interpreting the "selection effect" of biodiversity on ecosystem function. Ecology Letters 8:846–856.

France, K. E., and J. E. Duffy. 2006. Diversity and dispersal interactively affect predictability of ecosystem function. Nature 441:1139–1143.

Free, A., and N. H. Barton. 2007. Do evolution and ecology need the Gaia hypothesis? Trends in Ecology & Evolution 22:611–619.

Frost, T. M., S. R. Carpenter, A. I. Ives, and T. K. Kratz. 1995. Species compensation and complementarity in ecosystem function. Pages 224–239 in C. G. Jones and J. H. Lawton, eds. Linking species and ecosystems. New York, Chapman & Hall.

Fussmann, G. F., M. Loreau, and P. A. Abrams. 2007. Eco-evolutionary dynamics of communities and ecosystems. Functional Ecology 21:465–477.

Gardner, A., and S. A. West. 2004. Spite and the scale of competition. Journal of Evolutionary Biology 17:1195–1203.

Gardner, M. R., and W. R. Ashby. 1970. Connectance of large dynamic (cybernetic) systems: critical values for stability. Nature 228:784.

Gatto, M. 1990. A general minimum principle for competing populations: some ecological and evolutionary consequences. Theoretical Population Biology 37:369–388.

Gause, G. F. 1934. The struggle for existence. Baltimore, Williams & Wilkins.

Gause, G. F., and A. A. Wit. 1935. Behavior of mixed populations and the problem of natural selection. American Naturalist 64:596–609.

Goldberg, D. E., and A. M. Barton. 1992. Patterns and consequences of interspecific competition in natural communities: a review of field experiments with plants. American Naturalist 139:771–801.

Gonzalez, A., and E. J. Chaneton. 2002. Heterotroph species extinction, abundance and biomass dynamics in an experimentally fragmented microecosystem. Journal of Animal Ecology 71:594–602.

Gonzalez, A., and B. Descamps-Julien. 2004. Population and community variability in randomly fluctuating environments. Oikos 106:105–116.

Gonzalez, A., N. Mouquet, and M. Loreau. 2009. Biodiversity as spatial insurance: the effects of habitat fragmentation and dispersal on ecosystem functioning. Pages 134–146 *in* S. Naeem, D. E. Bunker, A. Hector, M. Loreau, and C. Perrings, eds. Biodiversity, ecosystem functioning, and human wellbeing: an ecological and economic perspective. Oxford, Oxford University Press.

Goreaud, F., M. Loreau, and C. Millier. 2002. Spatial structure and the survival of an inferior competitor: a theoretical model of neighbourhood competition in plants. Ecological Modelling 158:1–19.

Gotelli, N. J. 1991. Metapopulation models: the rescue effect, the propagule rain, and the core-satellite hypothesis. American Naturalist 138:768–776.

Gotelli, N. J., and G. R. Graves. 1996. Null models in ecology. Washington, Smithsonian Institution Press.

Gotelli, N. J., and W. G. Kelly. 1993. A general model of metapopulation dynamics. Oikos 68:36–44.

Gotelli, N. J., and D. J. McCabe. 2002. Species co-occurrence: a meta-analysis of J. M. Diamond's assembly rules model. Ecology 83:2091–2096.

Goudard, A., and M. Loreau. 2008. Nontrophic interactions, biodiversity, and ecosystem functioning: an interaction web model. American Naturalist 171:91–106.

Grace, J. B., T. M. Anderson, M. D. Smith, E. Seabloom, S. J. Andelman, G. Meche, E. Weiher, L. K. Allain, H. Jutila, M. Sankaran, J. Knops, M. Ritchie, and M. R. Willig. 2007. Does species diversity limit productivity in natural grassland communities? Ecology Letters 10:680–689.

Grime, J. P. 1979. Plant strategies and vegetation processes. New York, John Wiley & Sons.

Gross, K., and B. J. Cardinale. 2005. The functional consequences of random vs. ordered species extinctions. Ecology Letters 8:409–418.

Gross, K. L., M. R. Willig, L. Gough, R. Inouye, and S. B. Cox. 2000. Patterns of species density and productivity at different spatial scales in herbaceous plant communities. Oikos 89:417–427.

Grover, J. P. 1994. Assembly rules for communities of nutrient-limited plants and specialist herbivores. American Naturalist 143:258–282.

———. 1997. Resource competition. London, Chapman & Hall.

Gurevitch, J., L. L. Morrow, A. Wallace, and J. S. Walsh. 1992. A meta-analysis of competition in field experiments. American Naturalist 140:539–572.

Gustafson, E. J., and R. H. Gardner. 1996. The effect of landscape heterogeneity on the probability of patch colonization. Ecology 77:94–107.

Haegeman, B., and M. Loreau. 2008. Limitations of entropy maximization in ecology. Oikos 117:1700–1710.

———. 2009. Trivial and non-trivial applications of entropy maximization in ecology: a reply to Shipley. Oikos 118:1270–1278.

Hairston, N. G., Jr., F. E. Smith, and L. B. Slobodkin. 1960. Community structure, population control, and competition. American Naturalist 94:421–425.

Hamilton, W. D. 1964a. The genetical evolution of social behaviour. I. Journal of Theoretical Biology 7:1–16.

Hamilton, W. D. 1964b. The genetical evolution of social behaviour. II. Journal of Theoretical Biology 7:17–52.

Hanski, I. 1981. Coexistence of competitors in patchy environment with and without predation. Oikos 37:306–312.

———. 1983. Coexistence of competitors in patchy environment. Ecology 64: 493–500.

Hanski, I., and M. E. Gilpin. 1997. Metapopulation biology: ecology, genetics, and evolution. San Diego, California, Academic Press.

Hardin, G. 1960. The competitive exclusion principle. Science 131:1292–1297.

Harrison, G. W. 1979. Stability under environmental stress: resistance, resilience, persistence, and variability. American Naturalist 113:659–669.

Harte, J., T. Zillio, E. Conlisk, and A. B. Smith. 2008. Maximum entropy and the state variable approach to macroecology. Ecology 89:2700–2711.

Hastings, A. 1980. Disturbance, coexistence, history, and competition for space. Theoretical Population Biology 18:363–373.

Hättenschwiler, S., and P. Gasser. 2005. Soil animals alter plant litter diversity effects on decomposition. Proceedings of the National Academy of Sciences of the USA 102:1519–1524.

Heal, G. M., E. B. Barbier, K. J. Boyle, A. P. Covich, S. P. Gloss, C. H. Hershner, J. P. Hoehn, C. M. Pringle, S. Polasky, K. Segerson, and K. Shrader-Frechette. 2005. Valuing ecosystem services: toward better environmental decision making. Washington, D.C., National Academies Press.

Hector, A., and R. Bagchi. 2007. Biodiversity and ecosystem multifunctionality. Nature 448:188–190.

Hector, A., A. J. Beale, A. Minns, S. J. Otway, and J. H. Lawton. 2000. Consequences of the reduction of plant diversity and litter decomposition: effects through litter quality and microenvironment. Oikos 90:357–371.

Hector, A., M. Loreau, B. Schmid, and BIODEPTH. 2002. Biodiversity manipulation experiments: studies replicated at multiple sites. Pages 36–46 in M. Loreau, S. Naeem, and P. Inchausti, eds. Biodiversity and ecosystem functioning: synthesis and perspectives. Oxford, Oxford University Press.

Hector, A., B. Schmid, C. Beierkuhnlein, M. C. Caldeira, M. Diemer, P. G. Dimitrakopoulos, J. A. Finn, H. Freitas, P. S. Giller, J. Good, R. Harris, P. Högberg, K. Huss-Danell, J. Joshi, A. Jumpponen, C. Körner, P. W. Leadley, M. Loreau, A. Minns, C.P.H. Mulder, G. O'Donovan, S. J. Otway, J. S. Pereira, A. Prinz, D. J. Read, M. Scherer-Lorenzen, E.-D. Schulze, A.-S. D. Siamantziouras, E. M. Spehn, A. C. Terry, A. Y. Troumbis, F. I. Woodward, S. Yachi, and J. H. Lawton. 1999. Plant diversity and productivity experiments in European grasslands. Science 286:1123–1127.

Hedin, L. O., J. J. Armesto, and A. H. Johnson. 1995. Patterns of nutrient loss from unpolluted, old-growth temperate forests: evaluation of biogeochemical theory. Ecology 76:493–509.

Heemsbergen, D. A., M. P. Berg, M. Loreau, J. R. van Haj, J. H. Faber, and H. A. Verhoef. 2004. Biodiversity effects on soil processes explained by interspecific functional dissimilarity. Science 306:1019–1020.

Hilbert, D. W., D. M. Swift, J. K. Detling, and M. I. Dyer. 1981. Relative growth rates and the grazing optimization hypothesis. Oecologia 51:14–18.

Hillebrand, H., and B. J. Cardinale. 2004. Consumer effects decline with prey diversity. Ecology Letters 7:192–201.

Hobbie, S. E. 1992. Effects of plant species on nutrient cycling. Trends in Ecology and Evolution 7:336–339.

Holling, C. S. 1959. Some characteristics of simple types of predation and parasitism. Canadian Entomologist:385–398.

———. 1973. Resilience and stability of ecological systems. Annual Review of Ecology and Systematics 4:1–23.

Holt, R. D. 1977. Predation, apparent competition, and the structure of prey communities. Theoretical Population Biology 12:197–229.

———. 1995. Linking species and ecosystems: where's Darwin? Pages 273–279 *in* C. G. Jones and J. H. Lawton, eds. Linking species and ecosystems. New York, Chapman & Hall.

———. 2002. Food webs in space: On the interplay of dynamic instability and spatial processes. Ecological Research 17:261–273.

Holt, R. D., J. P. Grover, and D. Tilman. 1994. Simple rules for interspecific dominance in systems with exploitative and apparent competition. American Naturalist 144:741–771.

Holt, R. D., and M. Loreau. 2001. Biodiversity and ecosystem functioning: the role of trophic interactions and the importance of system openness. Pages 246–262 *in* A. P. Kinzig, S. W. Pacala, and D. Tilman, eds. The functional consequences of biodiversity: empirical progress and theoretical extensions. Princeton, New Jersey, Princeton University Press.

Holyoak, M., M. A. Leibold, and R. D. Holt. 2005. Metacommunities: spatial dynamics and ecological communities. Chicago, University of Chicago Press.

Holyoak, M., and M. Loreau. 2006. Reconciling empirical ecology with neutral community models. Ecology 87:1370–1377.

Hooper, D. U., F. S. Chapin, J. J. Ewel, A. Hector, P. Inchausti, S. Lavorel, J. H. Lawton, D. M. Lodge, M. Loreau, S. Naeem, B. Schmid, H. Setälä, A. J. Symstad, J. Vandermeer, and D. A. Wardle. 2005. Effects of biodiversity on ecosystem functioning: a consensus of current knowledge. Ecological Monographs 75:3–35.

Horn, H. S., and R. H. MacArthur. 1972. Competition among fugitive species in a harlequin environment. Ecology 53:749–752.

Houlahan, J. E., D. J. Currie, K. Cottenie, G. S. Cumming, S. K. M. Ernest, C. S. Findlay, S. D. Fuhlendorf, U. Gaedke, P. Legendre, J. J. Magnuson, B. H. McArdle, E. H. Muldavin, D. Noble, R. Russell, R. D. Stevens, T. J. Willis, I. P. Woiwod, and S. M. Wondzell. 2007. Compensatory dynamics are rare in natural ecological communities. Proceedings of the National Academy of Sciences of the USA 104:3273–3277.

Hsu, S. B., S. Hubbell, and P. Waltman. 1977. A mathematical theory for single nutrient competition in continuous cultures of microorganisms. SIAM Journal of Applied Mathematics 32:366–383.

Hubbell, S. P. 1979. Tree dispersion, abundance, and diversity in a tropical dry forest. Science 203:1299–1309.

———. 2001. The unified neutral theory of biodiversity and biogeography. Princeton, New Jersey, Princeton University Press.

Hubbell, S. P., and R. B. Foster. 1986. Biology, chance, and history and the structure of tropical rain forest tree communities. Pages 314–329 *in* J. Diamond and T. J. Case, eds. Community ecology. New York, Harper & Row.

Hughes, J. B., and J. Roughgarden. 1998. Aggregate community properties and the strength of species' interactions. Proceedings of the National Academy of Sciences of the USA 95:6837–6842.

———. 2000. Species diversity and biomass stability. American Naturalist 155: 618–627.

Huisman, J., and F. J. Weissing. 1999. Biodiversity of plankton by species oscillations and chaos. Nature 402:407–410.

Hurtt, G. C., and S. W. Pacala. 1995. The consequences of recruitment limitation: reconciling chance, history and competitive differences between plants. Journal of Theoretical Biology 176:1–12.

Huston, M. A. 1979. A general hypothesis of species diversity. American Naturalist 113:81–101.

———. 1997. Hidden treatments in ecological experiments: re-evaluating the ecosystem function of biodiversity. Oecologia 110:449–460.

Huston, M. A., D. DeAngelis, and W. Post. 1988. New computer models unify ecological theory. BioScience 38:682–691.

Huston, M. A., and D. L. DeAngelis. 1994. Competition and coexistence: the effects of resource transport and supply rates. American Naturalist 144:954–977.

Hutchinson, G. E. 1961. The paradox of the plankton. American Naturalist 95:137–145.

Ito, H. C., and T. Ikegami. 2006. Food-web formation with recursive evolutionary branching. Journal of Theoretical Biology 238:1–10.

Ives, A. R., B. J. Cardinale, and W. E. Snyder. 2005. A synthesis of subdisciplines: predator–prey interactions, and biodiversity and ecosystem functioning. Ecology Letters 8:102–116.

Ives, A. R., and S. R. Carpenter. 2007. Stability and diversity of ecosystems. Science 317:58–62.

Ives, A. R., K. Gross, and J. L. Klug. 1999. Stability and variability in competitive communities. Science 286:542–544.

Ives, A. R., and J. B. Hughes. 2002. General relationships between species diversity and stability in competitive systems. American Naturalist 159:388–395.

Ives, A. R., J. L. Klug, and K. Gross. 2000. Stability and species richness in complex communities. Ecology Letters 3:399–411.

Ives, A. R., S. T. Woody, E. V. Nordheim, C. Nelson, and J. H. Andrews. 2004. The synergistic effects of stochasticity and dispersal on population densities. American Naturalist 163:357–387.

Järemo, J., J. Tuomi, P. Nilsson, and T. Lennartsson. 1999. Plant adaptations to herbivory: mutualistic versus antagonistic coevolution. Oikos 84:313–320.

Jaynes, E. T. 1957. Information theory and statistical mechanics. Physical Review 106:620–630.

———. 2003. Probability theory: the logic of science. Cambridge, Cambridge University Press.

Jiang, L., H. Joshi, and S. N. Patel. 2009. Predation alters relationships between biodiversity and temporal stability. American Naturalist 173:389–399.

Johnson, C. R., and M. C. Boerlijst. 2002. Selection at the level of the community: the importance of spatial structure. Trends in Ecology and Evolution 17: 83–90.

Jones, C. G., and J. H. Lawton. 1995. Linking species and ecosystems. New York, Chapman & Hall.

Jones, C. G., J. H. Lawton, and M. Shachak. 1994. Organisms as ecosystem engineers. Oikos 69:373–386.

Keesing, F., R. D. Holt, and R. S. Ostfeld. 2006. Effects of species diversity on disease risk. Ecology Letters 9:485–498.

Kinzig, A. P., S. W. Pacala, and D. Tilman. 2001. The functional consequences of biodiversity: empirical progress and theoretical extensions. Princeton, New Jersey, Princeton University Press.

Kleiber, M. 1961. The fire of life: an introduction to animal energetics. New York, John Wiley & Sons.

Klug, J. L., J. M. Fischer, A. R. Ives, and B. Dennis. 2000. Compensatory dynamics in planktonic community responses to pH perturbations. Ecology 81:387–398.

Knight, T. M., M. W. McCoy, J. M. Chase, K. A. McCoy, and R. D. Holt. 2005. Trophic cascades across ecosystems. Nature 437:880–883.

Kokkoris, G. D., V. A. A. Jansen, M. Loreau, and A. Y. Troumbis. 2002. Variability in interaction strength and implications for biodiversity. Journal of Animal Ecology 71:362–371.

Kolasa, J., and B. L. Li. 2003. Removing the confounding effect of habitat specialization reveals the stabilizing contribution of diversity to species variability. Proceedings of the Royal Society of London B270:S198–S201.

Kondoh, M. 2003. Foraging adaptation and the relationship between food-web complexity and stability. Science 299:1388–1391.

Kooijman, S. A. L. M. 2000. Dynamic energy and mass budgets in biological systems. Cambridge, Cambridge University Press.

Kratina, P., M. Vos, and B. R. Anholt. 2007. Species diversity modulates predation. Ecology 88:1917–1923.

Kylafis, G., and M. Loreau. 2008. Ecological and evolutionary consequences of niche construction for its agent. Ecology Letters 11:1072–1081.

Lande, R. 1996. Statistics and partitioning of species diversity, and similarity among multiple communities. Oikos 76:5–13.

Lande, R., S. Engen, and B.-E. Saether. 2003. Stochastic population dynamics in ecology and conservation: Oxford series in ecology and evolution. Oxford, Oxford University Press.

Lata, J.-C., V. Degrange, X. Raynaud, P.-A. Maron, R. Lensi, and L. Abbadie. 2004. Grass populations control nitrification in savanna soils. Functional Ecology 18:605–611.

Lata, J.-C., K. Guillaume, V. Degrange, L. Abbadie, and R. Lensi. 2000. Relationships between root density of the African grass *Hyparrhenia diplandra* and nitrification at the decimetric scale: an inhibition-stimulation balance hypothesis. Proceedings of the Royal Society of London B267:595–600.

Law, R., and U. Dieckmann. 1998. Symbiosis through exploitation and the merger of lineages in evolution. Proceedings of the Royal Society of London B265: 1245–1253.

Lehman, C. L., and D. Tilman. 2000. Biodiversity, stability, and productivity in competitive communities. American Naturalist 156:534–552.

Leibold, M. 1989. Resource edibility and the effects of predators and productivity on the outcome of trophic interactions. American Naturalist 134:922–949.

Leibold, M. A. 1995. The niche concept revisited: mechanistic models and community context. Ecology 76:1371–1382.

———. 1998. Similarity and local co-existence of species in regional biotas. Evolutionary Ecology 12:95–110.

Leibold, M. A., M. Holyoak, N. Mouquet, P. Amarasekare, J. M. Chase, M. F. Hoopes, R. D. Holt, J. B. Shurin, R. Law, D. Tilman, M. Loreau, and A. Gonzalez. 2004. The metacommunity concept: a framework for multi-scale community ecology. Ecology Letters 7:601–613.

Lennartsson, T., J. Tuomi, and P. Nilsson. 1997. Evidence for an evolutionary history of overcompensation in the grassland biennial *Gentianella campestris* (Gentianaceae). American Naturalist 149:1147–1155.

Lenton, T. M. 1998. Gaia and natural selection. Nature 394:439–447.

Léon, J. A., and D. B. Tumpson. 1975. Competition between two species for two complementary or substitutable resources. Journal of Theoretical Biology 50:185–201.

Leroux, S. J., and M. Loreau. 2008. Subsidy hypothesis and strength of trophic cascades across ecosystems. Ecology Letters 11:1147–1156.

Levin, S. A. 1970. Community equilibria and stability, and an extension of the competitive exclusion principle. American Naturalist 104:413–423.

———. 1999. Fragile dominion: complexity and the commons. Reading, Massachusetts, Perseus Books.

Levine, S. H. 1976. Competitive interactions in ecosystems. American Naturalist 110:903–910.

Levins, R. 1968, Evolution in changing environments: some theoretical explorations. Princeton, New Jersey, Princeton University Press.

———. 1969. Some demographic and genetic consequences of environmental heterogeneity for biological control. Bulletin of the Entomological Society of America 15:237–240.

———. 1970. Complex systems. Pages 73–88 *in* C. H. Waddington, ed. Towards a theoretical biology. Edinburgh, Edinburgh University Press.

———. 1979. Coexistence in a variable environment. American Naturalist 114:765–783.

Levins, R., and D. Culver. 1971. Regional coexistence of species and competition between rare species. Proceedings of the National Academy of Science of the USA 68:1246–1248.

Lewontin, R. C. 1983. The organism as the subject and object of evolution. Scientia 118:65–83.

Lindeman, R. L. 1942. The trophic-dynamic aspect of ecology. Ecology 23:399–418.

Loeuille, N., and M. Loreau. 2005. Evolutionary emergence of size-structured food webs. Proceedings of the National Academy of Sciences of the USA. 102: 5761–5766.

———. 2010. Emergence of complex food-web structure in community evolution models. Pages 163–178 *in* H. A. Verhoef and P. J. Morin, eds. Community ecology: processes, models, and applications. Oxford, Oxford University Press.

Loreau, M. 1992. Time scale of resource dynamics and coexistence through time partitioning. Theoretical Population Biology 41:401–412.

———. 1994. Ground beetles in a changing environment: determinants of species diversity and community assembly. Pages 77–98 in T. J. B. Boyle and Christopher E. B., eds. Biodiversity, temperate ecosystems, and global change. Berlin, Springer-Verlag.

———. 1995. Consumers as maximizers of matter and energy flow in ecosystems. American Naturalist 145:22–42.

———. 1996. Coexistence of multiple food chains in a heterogeneous environment: interactions among community structure, ecosystem functioning, and nutrient dynamics. Mathematical Biosciences 134:153–188.

———. 1998a. Biodiversity and ecosystem functioning: a mechanistic model. Proceedings of the National Academy of Sciences of the USA 95:5632–5636.

———. 1998b. Ecosystem development explained by competition within and between material cycles. Proceedings of the Royal Society of London B265:33–38.

———. 1998c. Separating sampling and other effects in biodiversity experiments. Oikos 82:600–602.

———. 2000a. Are communities saturated? On the relationship between α, β, and γ diversity. Ecology Letters 3:73–76.

———. 2000b. Biodiversity and ecosystem functioning: recent theoretical advances. Oikos 91:3–17.

———. 2001. Microbial diversity, producer-decomposer interactions and ecosystem processes: a theoretical model. Proceedings of the Royal Society of London B268:303–309.

———. 2002. Evolutionary processes in ecosystems. Pages 292–297 in H. A. Mooney and J. Canadell, eds. Encyclopedia of Global Environmental Change, vol. 2, The Earth system: biological and ecological dimensions of global environmental change. London, John Wiley & Sons.

———. 2004. Does functional redundancy exist? Oikos 104:606–611.

———. 2008. Biodiversity and ecosystem functioning: the mystery of the deep sea. Current Biology 18:R126–R128.

———. 2010. The challenges of biodiversity science. Excellence in Ecology, Oldendorf/Luhe, International Ecology Institute (in press).

Loreau, M., and C. de Mazancourt. 2008. Species synchrony and its drivers: neutral and nonneutral community dynamics in fluctuating environments. American Naturalist 172:E48–E66.

Loreau, M., A. Downing, M. Emmerson, A. Gonzalez, J. Hughes, P. Inchausti, J. Joshi, J. Norberg, and O. Sala. 2002a. A new look at the relationship between diversity and stability. Pages 79–91 in M. Loreau, S. Naeem, and P. Inchausti, eds. Biodiversity and ecosystem functioning: synthesis and perspectives. Oxford, Oxford University Press.

Loreau, M., and A. Hector. 2001. Partitioning selection and complementarity in biodiversity experiments. Nature 412:72–76.

Loreau, M., and R. D. Holt. 2004. Spatial flows and the regulation of ecosystems. American Naturalist 163:606–615.

Loreau, M., and N. Mouquet. 1999. Immigration and the maintenance of local species diversity. American Naturalist 154:427–440.

Loreau, M., N. Mouquet, and A. Gonzalez. 2003a. Biodiversity as spatial insurance in heterogeneous landscapes. Proceedings of the National Academy of Sciences of the USA 100:12765–12770.

Loreau, M., N. Mouquet, and R. D. Holt. 2003b. Meta-ecosystems: a theoretical framework for a spatial ecosystem ecology. Ecology Letters 6:673–679.

———. 2005. From metacommunities to meta-ecosystems. Pages 418–438 *in* M. Holyoak, M. A. Leibold, and R. D. Holt, eds. Metacommunities: spatial dynamics and ecological communities. Chicago, University of Chicago Press.

Loreau, M., S. Naeem, and P. Inchausti. 2002b. Biodiversity and ecosystem functioning: synthesis and perspectives. Oxford, Oxford University Press.

Loreau, M., S. Naeem, P. Inchausti, J. Bengtsson, J. P. Grime, A. Hector, D. U. Hooper, M. A. Huston, D. Raffaelli, B. Schmid, D. Tilman, and D. A. Wardle. 2001. Biodiversity and ecosystem functioning: current knowledge and future challenges. Science 294:804–808.

Loreau, M., and E. Thébault. 2005. Food webs and the relationship between biodiversity and ecosystem functioning. Pages 270–282 *in* P. C. de Ruiter, V. Wolters, and J. C. Moore, eds. Dynamic food webs: multispecies assemblages, ecosystem development and environmental change. Burlington, Massachusetts, Academic Press.

Lotka, A. J. 1922. Contribution to the energetics of evolution. Proceedings of the National Academy of Sciences of the USA 8:147–151.

———. 1925. Elements of physical biology. Baltimore, Williams & Wilkins.

Lovelock, J. 1979. Gaia: a new look at life on Earth. Oxford, Oxford University Press.

MacArthur, R. H. 1955. Fluctuations of animal populations and a measure of community stability. Ecology 36:533–535.

———. 1969. Species packing, and what interspecies competition minimizes. Proceedings of the National Academy of Sciences of the USA 64:1369–1371.

———. 1970. Species packing and competitive equilibrium for many species. Theoretical Population Biology 1:1–11.

———. 1972. Geographical ecology: patterns in the distribution of species. Princeton, New Jersey, Princeton University Press.

MacArthur, R. H., and R. Levins. 1964. Competition, habitat selection, and character displacement in a patchy environment. Proceedings of the National Academy of Sciences of the USA 51:1207–1210.

———. 1967. The limiting similarity, convergence and divergence of coexistence species. American Naturalist 101:377–385.

MacArthur, R. H., and E. O. Wilson. 1967. The theory of island biogeography. Princeton, New Jersey, Princeton University Press.

Malthus, T. R. 1798. An essay on the principle of population. London, J. Johnson.

Margalef, R. 1963. On certain unifying principles in ecology. American Naturalist 97:357–374.

Matthiessen, B., and H. Hillebrand. 2006. Dispersal frequency affects local biomass production by controlling local diversity. Ecology Letters 9:652–662.

Maurer, B. A. 1999. Untangling ecological complexity: the macroscopic perspective. Chicago, Chicago University Press.

May, R. M. 1972. Will a large complex system be stable? Nature 238:413–414.

———. 1973. Stability and complexity in model ecosystems. Princeton, New Jersey, Princeton University Press.

———. 1974. Ecosystem patterns in randomly fluctuating environments. Pages 1–52 *in* R. Rosen and F. M. Snell, eds. Progress in Theoretical Biology. New York, Academic Press.

———. 1981. Models for two interacting populations. Pages 78–104 *in* R. M. May, ed. Theoretical ecology: principles and applications. Sunderland, Massachusetts, Sinauer.

McCann, K., A. Hastings, and G. R. Huxel. 1998. Weak trophic interactions and the balance of nature. Nature 395:794–798.

McCann, K. S. 2000. The diversity-stability debate. Nature 405:228–233.

McCann, K. S., J. B. Rasmussen, and J. Umbanhowar. 2005. The dynamics of spatially coupled food webs. Ecology Letters 8:513–523.

McGill, B. J., B. J. Enquist, E. Weiher, and M. Westoby. 2006a. Rebuilding community ecology from functional traits. Trends in Ecology and Evolution 21:178–185.

McGill, B. J., R. S. Etienne, J. S. Gray, D. Alonso, M. J. Anderson, H. K. Benecha, M. Dornelas, B. J. Enquist, J. L. Green, F. L. He, A. H. Hurlbert, A. E. Magurran, P. A. Marquet, B. A. Maurer, A. Ostling, C. U. Soykan, K. I. Ugland, and E. P. White. 2007. Species abundance distributions: moving beyond single prediction theories to integration within an ecological framework. Ecology Letters 10:995–1015.

McGill, B. J., B. A. Maurer, and M. D. Weiser. 2006b. Empirical evaluation of neutral theory. Ecology 87:1411–1423.

McKane, A. J. 2004. Evolving complex food webs. European Physical Journal B38:287–295.

McNaughton, S. J. 1977. Diversity and stability of ecological communities: a comment on the role of empiricism in ecology. American Naturalist 111:515–525.

———. 1979. Grazing as an optimization process: grass-ungulate relationships in the Serengeti. American Naturalist 113:691–703.

———. 1986. On plants and herbivores. American Naturalist 128:765–770.

———. 1993. Biodiversity and stability of grazing ecosystems. Pages 361–383 *in* E.-D. Schulze and H. A. Mooney, eds. Biodiversity and ecosystem function. Berlin, Springer-Verlag.

McQueen, D. J., J. R. Post, and E. L. Mills. 1986. Trophic relationship in freshwater pelagic ecosystems. Canadian Journal of Fisheries and Aquatic Sciences 43:1571–1581.

MEA. 2005. Ecosystems and human well-being: Synthesis, millennium ecosystem assessment. Washington, D.C., Island Press.

Metz, J. A. J., and O. Diekmann. 1986. The dynamics of physiologically structured populations: lecture notes in biomathematics. Berlin, Springer-Verlag.

Mittelbach, G. G., C. F. Steiner, S. M. Scheiner, K. L. Gross, H. L. Reynolds, R. B. Waide, M. R. Willig, S. I. Dodson, and L. Gouch. 2001. What is the observed relationship between species richness and productivity? Ecology 82:2381–2396.

Morowitz, H. J. 1966. Physical background of cycles in biological systems. Journal of Theoretical Biology 13:60–62.

Mouquet, N., and M. Loreau. 2002. Coexistence in metacommunities: The regional similarity hypothesis. American Naturalist 159:420–426.

———. 2003. Community patterns in source-sink metacommunities. American Naturalist 162:544–557.

Mulder, C. P. H., D. D. Uliassi, and D. F. Doak. 2001. Physical stress and diversity-productivity relationships: the role of positive interactions. Proceedings of the National Academy of Sciences of the USA 98:6704–6708.

Murrell, D. J., and R. Law. 2003. Heteromyopia and the spatial coexistence of similar competitors. Ecology Letters 6:48–59.

Naeem, S. 1998. Species redundancy and ecosystem reliability. Conservation Biology 12:39–45.

Naeem, S., D. E. Bunker, A. Hector, M. Loreau, and C. Perrings. 2009. Biodiversity, ecosystem functioning, and human wellbeing: an ecological and economic perspective. Oxford, Oxford University Press.

Naeem, S., D. R. Hahn, and G. Schuurman. 2000. Producer-decomposer codependency influences biodiversity effects. Nature 403:762–764.

Naeem, S., and J. P. Wright. 2003. Disentangling biodiversity effects on ecosystem functioning: deriving solutions to a seemingly insurmountable problem. Ecology Letters 6:567–579.

Nakano, S., and M. Murakami. 2001. Reciprocal subsidies: dynamics interdependence between terrestrial and aquatic food webs. Proceedings of the National Academy of Sciences of the USA 98:166–170.

Neubert, M. G., and H. Caswell. 1997. Alternatives to resilience for measuring the responses of ecological systems to perturbations. Ecology 78:653–665.

Neutel, A. M., J. A. P. Heesterbeek, J. van de Koppel, G. Hoenderboom, A. Vos, C. Kaldeway, F. Berendse, and P. C. de Ruiter. 2007. Reconciling complexity with stability in naturally assembling food webs. Nature 449:599–602.

Norberg, J., D. P. Swaney, J. Dushoff, J. Lin, R. Casagrandi, and S. A. Levin. 2001. Phenotypic diversity and ecosystem functioning in changing environments: a theoretical framework. Proceedings of the National Academy of Sciences of the USA 98:11376–11381.

O'Neill, R. V., D. L. DeAngelis, J. B. Waide, and T. F. H. Allen. 1986. A hierarchical concept of ecosystems. Princeton, New Jersey, Princeton University Press.

Odling-Smee, F. J., K. N. Laland, and M. W. Feldman. 2003. Niche construction: the neglected process in evolution. Princeton, New Jersey, Princeton University Press.

Odum, E. P. 1953. Fundamentals of ecology. Philadelphia, Saunders.

———. 1969. The strategy of ecosystem development. Science 164:262–270.

Odum, H. T. 1983. Systems ecology: an introduction. New York, John Wiley & Sons.

Odum, H. T., and R. C. Pinkerton. 1955. Times speed regulator: the optimum efficiency for maximum power output in physical and biological systems. American Scientist 43:331–343.

Oksanen, L., S. D. Fretwell, J. Arruda, and P. Niemela. 1981. Exploitation ecosystems in gradients of primary productivity. American Naturalist 118:240–261.

Ostfeld, R. S., and F. Keesing. 2000. Biodiversity and disease risk: the case of Lyme disease. Conservation Biology 14:722–728.

Ostfeld, R. S., and K. LoGiudice. 2003. Community disassembly, biodiversity loss, and the erosion of an ecosystem service. Ecology 84:1421–1427.

Owen, D. F., and R. G. Wiegert. 1976. Do consumers maximize plant fitness? Oikos 27:488–492.

———. 1981. Mutualism between grasses and grazers: an evolutionary hypothesis. Oikos 36:376–492.

Owen-Smith, N. 2002. A metaphysiological modelling approach to stability in herbivore-vegetation systems. Ecological Modelling 149:153–178.

Paige, K. N., and T. G. Whitham. 1987. Overcompensation in response to mammalian herbivory: the advantage of being eaten. American Naturalist 129: 407–416.

Park, T. 1948. Experimental studies of interspecies competition. I. Competition between populations of flour beetles, *Tribolium confusum* Duval and *Tribolium castaneum* Herbst. Ecological Monographs 18:265–308.

Patten, B. C. 1975. Ecosystem linearization: an evolutionary design problem. American Naturalist 109:529–539.

Perakis, S. S., and L. O. Hedin. 2002. Nitrogen loss from unpolluted South American forests mainly via dissolved organic compounds. Nature 415:416–419.

Petchey, O. L., and K. J. Gaston. 2002. Functional diversity (FD), species richness and community composition. Ecology Letters 5:402–411.

Peters, R. H. 1983. The ecological implications of body size: Cambridge studies in ecology. Cambridge, Cambridge University Press.

Petrusewicz, K., and A. Macfadyen. 1970. Productivity of terrestrial animals: principles and methods: IBP handbook. Philadelphia, F. A. Davis.

Pickett, S. T. A., and M. L. Cadenasso. 1995. Landscape ecology: spatial heterogeneity in ecological systems. Science 269:331–334.

Pimm, S. 1982. Food webs. London, Chapman & Hall.

Pimm, S. L. 1984. The complexity and stability of ecosystems. Nature 307:321–326.

———. 1991. The balance of nature? Ecological issues in the conservation of species and communities. Chicago, University of Chicago Press.

Pimm, S. L., and J. H. Lawton. 1977. Number of trophic levels in ecological communities. Nature 268:329–331.

Polis, G. A., W. B. Anderson, and R. D. Holt. 1997. Toward an integration of landscape and food web ecology: the dynamics of spatially subsidized food webs. Annual Review of Ecology and Systematics 28:289–316.

Price, G. R. 1995. The nature of selection. Journal of Theoretical Biology 175:389–396.

Puccia, C. J., and R. Levins. 1985. Qualitative modeling of complex systems: an introduction to loop analysis and time averaging. Cambridge, Massachusetts, Harvard University Press.

Ricklefs, R. E., and D. Schluter. 1993. Species diversity: regional and historical influences. Pages 350–363 *in* R. E. Ricklefs and D. Schluter, eds. Species diversity in ecological communities: historical and geographical perspectives. Chicago, University of Chicago Press.

Rixen, C., and C. P. H. Mulder. 2005. Improved water retention links high species richness with increased productivity in arctic tundra moss communities. Oecologia 146:287–299.

Romanuk, T. N., and J. Kolasa. 2004. Population variability is lower in diverse rock pools when the obscuring effects of local processes are removed. Ecoscience 11:455–462.

Rooney, N., K. McCann, G. Gellner, and J. C. Moore. 2006. Structural asymmetry and the stability of diverse food webs. Nature 442:265–269.

Roscher, C., V. M. Temperton, M. Scherer-Lorenzen, M. Schmitz, J. Schumacher, B. Schmid, N. Buchmann, W. W. Weisser, and E.-D. Schulze. 2005. Overyielding in experimental grassland communities—irrespective of species pool or spatial scale. Ecology Letters 8:576–577.

Rosenzweig, M. L. 1971. Paradox of enrichment: destabilization of exploitation ecosystems in ecological time. Science 171:385–387.

———. 1973. Evolution of the predator isocline. Evolution 27:84–94.

Rosenzweig, M. L., and R. H. MacArthur. 1963. Graphical representation and stability conditions of predator-prey interactions. American Naturalist 97:209–223.

Rossberg, A. G., H. Matsuda, T. Amemiya and K. Itoh. 2006. Some properties of the speciation model for food-web structure: mechanisms for degree distributions and intervality. Journal of Theoretical Biology 238:401–415.

Savage, V. M., C. T. Webb, and J. Norberg. 2007. A general multi-trait-based framework for studying the effects of biodiversity on ecosystem functioning. Journal of Theoretical Biology 247:213–229.

Schmitz, O. J. 2007. Predator diversity and trophic interactions. Ecology 88:2415–2426.

Schmitz, O. J., V. Krivan, and O. Ovadia. 2004. Trophic cascades: the primacy of trait-mediated indirect interactions. Ecology Letters 7:153–163.

Schneider, E. D., and J. J. Kay. 1994. Life as a manifestation of the second law of thermodynamics. Mathematical and Computer Modelling 19:25–48.

Schoener, T. W. 1983a. Field experiments on interspecific competition. American Naturalist 122:240–285.

———. 1983b. Rate of species turnover decreases from lower to higher organisms: a review of the data. Oikos 41:372–377.

———. 1986. Mechanistic approaches to community ecology: a new reductionism? American Zoologist 26:81–106.

Schulze, E.-D., and H. A. Mooney. 1993. Biodiversity and ecosystem function. Berlin, Springer-Verlag.

Shmida, A., and M. V. Wilson. 1985. Biological determinants of species diversity. Journal of Biogeography 12:1–20.

Shurin, J. B., E. T. Borer, E. W. Seabloom, K. Anderson, C. A. Blanchette, B. Broitman, S. D. Cooper, and B. Halperm. 2002. A cross-ecosystem comparison of the strength of trophic cascades. Ecology Letters 5:785–791.

Silvertown, J. 1982. No evolved mutualism between grasses and grazers. Oikos 38:253–259.

Slatkin, M. 1974. Competition and regional coexistence. Ecology 55:128–134.

Sober, E., and D. S. Wilson. 1997. Unto others: the evolution of altruism. Cambridge, Massachusetts, Harvard University Press.

Solé, R. V., and J. Bascompte. 2006. Self-organization in complex ecosystems. Princeton, New Jersey, Princeton University Press.

Sommer, U. 1985. Comparison between steady state and non-steady state competition: experiments with natural phytoplankton. Limnology and Oceanography 30:335–346.

Spehn, E. M., A. Hector, J. Joshi, M. Scherer-Lorenzen, B. Schmid, E. Bazeley-White, C. Beierkuhnlein, M. C. Caldeira, M. Diemer, P. G. Dimitrakopoulos, J. A. Finn, H. Freitas, P. S. Giller, J. Good, R. Harris, P. Högberg, K. Huss-Danell, A. Jumpponen, J. Koricheva, P. W. Leadley, M. Loreau, A. Minns, C.P.H. Mulder, G. O'Donovan, S. J. Otway, C. Palmborg, J. S. Pereira, A. B. Pfisterer, A. Prinz, D. J. Read, E.-D. Schulze, A.-S. D. Siamantziouras, A. C. Terry, A. Y. Troumbis, F. I. Woodward, S. Yachi, and J. H. Lawton. 2005. Ecosystem effects of biodiversity manipulations in European grasslands. Ecological Monographs 75:37–63.

Steiner, C. F., Z. T. Long, J. A. Krumins, and P. J. Morin. 2005. Temporal stability of aquatic food webs: partitioning the effects of species diversity, species composition and enrichment. Ecology Letters 8:819–828.

Sterner, R. W., and J. J. Elser. 2002. Ecological stoichiometry: the biology of elements from molecules to the biosphere. Princeton, New Jersey, Princeton University Press.

Strong, D. R., Jr., D. Simberloff, L. G. Abele, and A. B. Thistle. 1984. Ecological communities: conceptual issues and the evidence. Princeton, New Jersey, Princeton University Press.

Suding, K. N., S. Lavorel, F. S. Chapin III, J. H. C. Cornelissen, S. Diaz, E. Garnier, D. Goldberg, D. U. Hooper, S. T. Jackson, and M.-L. Navas. 2008. Scaling environmental change through the community-level: a trait-based response-and-effect framework for plants. Global Change Biology 14:1125–1140.

Switzer, G. L., and L. E. Nelson. 1972. Nutrient accumulation and cycling in loblolly pine (*Pinus taeda* L.) plantation ecosystems: the first twenty years. Soil Science Society of America Journal 36:143–147.

Terborgh, J. W., and J. Faaborg. 1980. Saturation of bird communities in the West Indies. American Naturalist 116:178–195.

Thébault, E., and M. Loreau. 2003. Food-web constraints on biodiversity-ecosystem functioning relationships. Proceedings of the National Academy of Sciences of the USA 100:14949–14954.

———. 2005. Trophic interactions and the relationship between species diversity and ecosystem stability. American Naturalist 166:E95–E114.

———. 2006. The relationship between biodiversity and ecosystem functioning in food webs. Ecological Research 21:17–25.

Tilman, D. 1980. Resources: a graphical-mechanistic approach to competition and predation. American Naturalist 116:362–393.

———. 1982. Resource competition and community structure. Princeton, New Jersey, Princeton University Press.

———. 1988. Plant strategies and the dynamics and structure of plant communities. Princeton, New Jersey, Princeton University Press.

———. 1994. Competition and biodiversity in spatially structured habitats. Ecology 75:2–16.

———. 1996. Biodiversity: population versus ecosystem stability. Ecology 77:350–363.

Tilman, D. 1997. Distinguishing between the effects of species diversity and species composition. Oikos 80:185.

———. 1999. The ecological consequences of changes in biodiversity: a search for general principles. Ecology 80:1455–1474.

Tilman, D., and P. Kareiva. 1997. Spatial ecology: the role of space in population dynamics and interspecific interactions. Princeton, New Jersey, Princeton University Press.

Tilman, D., J. Knops, D. Wedin, P. Reich, M. Ritchie, and E. Siemann. 1997a. The influence of functional diversity and composition on ecosystem processes. Science 277:1300–1302.

Tilman, D., C. L. Lehman, and C. E. Bristow. 1998. Diversity-stability relationships: statistical inevitability or ecological consequence? American Naturalist 151:277–282.

Tilman, D., C. L. Lehman, and K. T. Thomson. 1997b. Plant diversity and ecosystem productivity: theoretical considerations. Proceedings of the National Academy of Sciences of the USA 94:1857–1861.

Tilman, D., S. Naeem, J. Knops, P. Reich, E. Siemann, D. Wedin, M. Ritchie, and J. Lawton. 1997c. Biodiversity and ecosystem properties. Science 278:1865–1869.

Tilman, D., P. B. Reich, J. Knops, D. Wedin, T. Mielke, and C. Lehman. 2001. Diversity and productivity in a long-term grassland experiment. Science 294:843–845.

Tilman, D., P. B. Reich, and J. M. H. Knops. 2006. Biodiversity and ecosystem stability in a decade-long grassland experiment. Nature 441:629–632.

Tilman, D., and D. Wedin. 1991. Plant traits and resource reduction for five grasses growing on a nitrogen gradient. Ecology 72:685–700.

Trenbath, B. R. 1974. Biomass productivity of mixtures. Advances in Agronomy 26:177–210.

Turner, M. G. 1989. Landscape ecology: the effect of pattern on process. Annual Review of Ecology and Systematics 20:171–197.

Turner, M. G., R. H. Gardner, and R. V. O'Neill. 2001. Landscape ecology in theory and practice: pattern and process. New York, Springer.

Ulanowicz, R. E. 1990. Aristotelean causalities in ecosystem development. Oikos 57:42–48.

Ulanowicz, R. E., and B. M. Hannon. 1987. Life and the production of entropy. Proceedings of the Royal Society of London B232:181–192.

Valone, T. J., and C. D. Hoffman. 2003. Population stability is higher in more diverse annual plant communities. Ecology Letters 6:90–95.

van Ruijven, J., and F. Berendse. 2003. Positive effects of plant species diversity on productivity in the absence of legumes. Ecology Letters 6:170–175.

———. 2005. Diversity-productivity relationships: Initial effects, long-term patterns, and underlying mechanisms. Proceedings of the National Academy of Sciences of the USA 102:695–700.

Vandermeer, J. 1989. The ecology of intercropping. Cambridge, Cambridge University Press.

Vasseur, D. A., and J. W. Fox. 2007. Environmental fluctuations can stabilize food web dynamics by increasing synchrony. Ecology Letters 10:1066–1074.

Venail, P. A., R. C. MacLean, T. Bouvier, M. A. Brockhurst, M. E. Hochberg, and N. Mouquet. 2008. Diversity and productivity peak at intermediate dispersal rate in evolving metacommunities. Nature 452:210–214.

Verhulst, P. F. 1838. Notice sur la loi que la population poursuit dans son accroissement. Correspondance mathématique et physique 10:113–121.

Vila, M., J. Vayreda, L. Comas, J. J. Ibanez, T. Mata, and B. Obon. 2007. Species richness and wood production: a positive association in Mediterranean forests. Ecology Letters 10:241–250.

Vitousek, P. M. 2004. Nutrient cycling and limitation: Hawaii as a model system. Princeton, New Jersey, Princeton University Press.

Vitousek, P. M., and P. A. Matson. 2009. Nutrient cycling and biogeochemistry. Pages 330–339 in S. A. Levin, S. R. Carpenter, H. C. J. Godfray, A. P. Kinzig, M. Loreau, J. B. Losos, B. Walker, and D. S. Wilcove, eds. Princeton guide to ecology. Princeton, New Jersey, Princeton University Press.

Vitousek, P. M., H. A. Mooney, J. Lubchenco, and J. M. Melillo. 1997. Human domination of Earth's ecosystems. Science 277:494–499.

Vitousek, P. M., and W. A. Reiners. 1975. Ecosystem succession and nutrient retention: a hypothesis. BioScience 25:376–381.

Vogt, R. J., T. N. Romanuk, and J. Kolasa. 2006. Species richness-variability relationships in multi-trophic aquatic microcosms. Oikos 113:55–66.

Volterra, V. 1926. Variazioni e fluttuazioni del numero d'individui in specie animali conviventi. Memorie della Reale Accademia dei Lincei 2:31–113.

Waide, R. B., M. R. Willig, C. F. Steiner, G. Mittelbach, L. Gough, S. I. Dodson, G. P. Juday, and R. Parmenter. 1999. The relationship between productivity and species richness. Annual Review of Ecology and Systematics 30:257–300.

Walker, L. R., and F. S. Chapin, III. 1987. Interactions among processes controlling successional change. Oikos 50:131–135.

Walker, T. W., and J. K. Syers. 1976. The fate of phosphorus during pedogenesis. Geoderma 15:1–19.

Wardle, D. A. 1999. Is "sampling effect" a problem for experiments investigating biodiversity-ecosystem function relationships? Oikos 87:403–407.

Wardle, D. A., L. R. Walker, and R. D. Bardgett. 2004. Ecosystem properties and forest decline in contrasting long-term chronosequences. Science 305:509–513.

Wardle, D. A., O. Zackrisson, G. Hörnberg, and C. Gallet. 1997. The influence of island area on ecosystem properties. Science 277:1296–1299.

Watson, A. J., and J. E. Lovelock. 1983. Biological homeostasis of the global environment: the parable of Daisyworld. Tellus 35B:284–289.

Webb, C. 2003. A complete classification of Darwinian extinction in ecological interactions. American Naturalist 161:181–205.

Webster, J. R., J. B. Waide, and B. C. Patten. 1974. Nutrient recycling and the stability of ecosystems. Pages 1–27 in F. G. Horwell, J. B. Gentry, and M. H. Smith, eds. Mineral cycling in southeastern ecosystems. Springfield, Virginia, National Technical Information Service.

Werner, E. E., and S. D. Peacor. 2003. A review of trait-mediated indirect interactions in ecological communities. Ecology 84:1083–1100.

Whitham, T. G., J. K. Bailey, J. A. Schweitzer, S. M. Shuster, R. K. Bangert, C. J. Leroy, E. V. Lonsdorf, G. J. Allan, S. P. DiFazio, B. M. Potts, D. G. Fischer,

C. A. Gehring, R. L. Lindroth, J. C. Marks, S. C. Hart, G. M. Wimp, and S. C. Wooley. 2006. A framework for community and ecosystem genetics: from genes to ecosystems. Nature Reviews Genetics 7:510–523.

Whittaker, R. H. 1972. Evolution and measurement of species diversity. Taxon 21:213–251.

Williams, H. T. P., and T. M. Lenton. 2007. The Flask model: emergence of nutrient-recycling microbial ecosystems and their disruption by environment-altering "rebel" organisms. Oikos 116:1087–1105.

Williams, R. J., and N. D. Martinez. 2000. Simple rules yield complex food webs. Nature 404:180–183.

Wilson, D. S. 1980. The natural selection of populations and communities. Menlo Park, California, Benjamin/Cummings.

———. 1997. Altruism and organism: disentangling the themes of multilevel selection theory. American Naturalist 150:S122–S134.

Wilson, D. S., and E. Sober. 1989. Reviving the superorganism. Journal of Theoretical Biology 136:337–356.

Wilson, E. O. 1992. The diversity of life. New York, W. W. Norton.

———. 2002. The future of life. New York, Vintage Books.

With, K. A. 1997. The application of neutral landscape models in conservation biology. Conservation Biology 11:1069–1080.

Wood, T. 1984. Phosphorus cycling in a northern hardwood forest: biological and chemical control. Science 223:391–393.

Wright, J. P., S. Naeem, A. Hector, C. Lehman, P. B. Reich, B. Schmid, and D. Tilman. 2006. Conventional functional classification schemes underestimate the relationship with ecosystem functioning. Ecology Letters 9:111–120.

Yachi, S., and M. Loreau. 1999. Biodiversity and ecosystem productivity in a fluctuating environment: the insurance hypothesis. Proceedings of the National Academy of Sciences of the USA 96:1463–1468.

Yodzis, P., and S. Innes. 1992. Body size and consumer-resource dynamics. American Naturalist 139:1151–1175.

Index

MONOGRAPHS IN POPULATION BIOLOGY
EDITED BY SIMON A. LEVIN AND HENRY S. HORN

Milton Keynes UK
Ingram Content Group UK Ltd.
UKHW020820230924
448618UK00020B/178